记录自然之美
To Track the Beauty of Nature

IBE

传递生命感动
To Convey the Sensation of Life

10

2009—2018

献给参加 IBE 工作及给予 IBE 帮助支持的人们

中国自然影像志

NATURE AND WILDLIFE PHOTOGRAPHY IN CHINA

影像生物调查所（IBE） 编

北京出版集团公司
北 京 出 版 社

图书在版编目（CIP）数据

中国自然影像志 / 影像生物调查所（IBE）编. —北京：北京出版社，2019.4
ISBN 978-7-200-12626-6

Ⅰ. ①中… Ⅱ. ①影… Ⅲ. ①珍稀动物—中国—摄影集②珍稀植物—中国—摄影集 Ⅳ. ①Q95-49 ②Q94-49

中国版本图书馆CIP数据核字（2016）第284631号

项目策划：安　东　王嶲灵
项目统筹：宋佩谦　司徒剑萍
责任编辑：王嶲灵　宋佩谦
装帧设计：李珊珊　吕丽梅　赵赟普
责任印制：武绽蕾　承伯平

中国自然影像志
ZHONGGUO ZIRAN YINGXIANG ZHI
影像生物调查所（IBE）　编

出　版　北京出版集团公司
　　　　北京出版社
地　址　北京北三环中路6号
邮　编　100120
网　址　www.bph.com.cn
总 发 行　北京出版集团公司
版 印 次　2019年4月第1版第1次印刷
印　刷　北京雅昌艺术印刷有限公司
开　本　787毫米×1092毫米　1/16
印　张　19.375
字　数　160千字
书　号　ISBN 978-7-200-12626-6
定　价　298.00元
如有印装质量问题，由本社负责调换
质量监督电话：010-58572393

《中国自然影像志》
编委会

contents

01 序言 II
Preface

02 地理分区 1
Ground partition

03 哺乳动物 31
Mammals

04 鸟类 117
Birds

05 植物 183
Plants

06 昆虫和蜘蛛 225
Insects and Spiders

07 两爬和鱼类 255
Amphibans, reptiles and fish

08 水下生物 277
Underwater creatures

09 影像生物调查所（IBE）简介 282
Introduction to IBE

10 后记 283
Afterword

序言
Preface

4 月末的一个清晨，天刚蒙蒙亮，我起身离开宾馆，前往不远处的北京奥林匹克森林公园。虽然天色尚早，公园里却已有不少散步、跑步的人，只有我带着双筒望远镜。正值鸟类大量迁徙的季节，很适合在工作前来这里待上几个小时。林间，柳莺和捕蝇鸟的叫声不绝于耳。

一转弯，我突然看见二十几个人紧紧地挤在一起，端着相机和“长枪短炮”，对准了一小片开阔地，好像在等着什么。就算是大牌明星到场，恐怕粉丝也不会如此专注，更别说用上这么昂贵的专业相机和器材。我站在他们身后，想一问究竟。其中一人骄傲地向我展示了他早前拍摄的一张照片：一只雄雉。哦，原来大家都在等着它呢。

在之后的几周里，每每清晨前来，都会遇见几拨同样热切的野生生物摄影师。湖边，每隔一段就能看见有人端着相机，耐心地等待着鸟儿出现。同样受到关注的还有一群在这里歇脚、准备继续飞往俄罗斯北部的大天鹅。

我之前当然也遇见过其他的野生生物摄影师，但北京的人数之多、专注度之高，无不说明中国人对此的热情。《中国自然影像志》就是一个很好的例证，其中收录了大量哺乳动物、鸟类、花草、昆虫和山水的照片，令人心醉。但中国摄影师的志向还远不止于此。

2009 年，这群摄影师组建了“影像生物调查所”，利用视像与野外考察相结合的方法，记录下包括中国 60 个自然保护区在内的不同地区的生物多样性编目。他们还与科学团队合作，以确保自己对生物种类的判断无误。这本书就是他们在过去 10 年的时间里，行走 30 多万千米，记录下 8000 种野生生物物种的精华。

中国的野生生物分布于种类繁多的栖息地内。书中，黑颈鹤俯身冲下大雪覆盖的喜马拉雅山顶，藏野驴和普氏野马漫步在广阔的草地，大象、猴子和鸟儿出没于潮湿的热带森林……此外还有长江江豚、热带珊瑚等。中国地域辽阔，从极寒之地到炎热地区，从干旱的沙漠到潮湿的森林，有着各种各样的物种。在这片与美国本土大陆面积相当的土地上，生活着世界上 15% 的陆栖脊椎动物以及 12% 的植物。在这里，野生生物摄影师有如此多的物种可以拍摄！

而通过植物摄影，我们看到雪融化后的几周里高山上绽放的清丽、雨林中的兰花、干旱沙漠里的生命、湖面上漂浮着的淡雅。每一朵花就像一颗宝石，让大地缀满了色彩。

本书的摄影师们得端着巨大的长焦镜头去接近动物。那当然，否则怎么可能拍到一对雪豹？但除此之外，他们还运用到了其他很多技术。特写镜头下呈现出一个昆虫的世界，精彩得就连科幻小说家都难以想象。

红外线触发相机捕捉到的一系列照片也同样令我惊喜。它们或许不具备传统照片的艺术性，但科学意义重大，并且代表着重要的技术进步。就在十几年前，数码相机的反应还很慢，动物走过面前时根本来不及拍摄。摄影机虽然能够拍摄，但必须不时地去查看和调整。如今，数码相机可以耐心地隐藏数月，抓拍上千张照片。它们捕捉到了我们知之甚少的一些动物，这些物种非常惧怕人类，很难用其他方式观察到。现在，我们终于开始了解那些隐藏在我们视线以外的生命了。

翻看这本书时，我不断地感受到惊喜。例如，书中用 10 多页介绍了几种不同的雉科动物，一种比一种美。我想看这个！不，那个！中国一共有 35 种雉类，是世界上雉类品种最多的国家。它们居住在中国西部和南部的山间，用那里的自然景色衬托出自己的靓丽。书里还有藏羚、川金丝猴，如果有幸遇见它们，一定要细细看上几个钟头。高原草甸上盛开着鲜艳的花，登山时一定会忍不住驻足观望。

而在这一切美好的背后，藏着我们深深的担忧，不知道后代是否还能和我们一样，欣赏并拍到这神奇的大自然。有一些物种已经非常稀少。最著名的例子就是大熊猫。它们曾经遍布中国大地，但现在基本只能在四川的山区找到了。更糟糕的是，大熊猫呈一个个孤立的族群分布，有一些族群因为过小而无法长久生存。此外，云南的热带森林里，人们为开展橡胶种植而占用了大片的土地。长江江豚也在沿岸工业活动的影响下走向灭绝，而全国近 7.5 亿人口居住在这一流域。

近年的研究证明，中国特有的物种——世界其他地区都没有——大多集中于西部地区。保护它们也就是保护中国的生物多样性。当然，这里还住着中国官方重点保护的对象——大熊猫。

于是我不禁发问：保护大熊猫，对于保护其他物种是否也有利？我和中国同事李彬彬共同研究后得出结论：有许多中国特有的物种与大熊猫生活在同一片地区，在大熊猫的“保护伞”下得以生存。当然，这并不是一个完美的方案，因为还有许多物种生活在其他地方，也需要保护。我们热爱大熊猫，但我们也同样热爱本书里拍摄到的其他物种。

这就是本书的意义所在。中国的野生世界不仅美得令人神往，更具有重要的科学价值。这些自然风光以其独特的魅力，吸引着来自国内和国外的游客。而本书则告诉我们，如果不着力保护自然，我们将会失去什么。

所以，当遇见那群等待雄雉的摄影师时，我很高兴。我当时就应该告诉他们，看见他们这么喜爱这种可爱的鸟类，我有多激动。正因为有了这份热情，我们才可以做出承诺，为了子孙后代、为了中国未来数千年的文化传承，保护好我们丰富的野生动植物资源。

斯图尔特·皮姆
杜克大学尼古拉斯环境学院教授

Preface

At sunrise, one late April morning, I walked the short distance from my hotel to the Olympic Forest Park in Beijing. Even at this early hour, many were walking and running its trails, but I was there with my binoculars. At peak bird migration, this was a great place to spend a couple of hours before work. The woods and marshes were busy with songs of warblers and flycatchers.

Coming round a bend in a path, a group of twenty people surprised me. Huddled close together, they stood with cameras and huge telephoto lenses looking across a small open field with great anticipation. No famous film star could have expected a more attentive group of fans, nor a more expensive collection of professional cameras and equipment. I stood at the back of the group, hoping for some explanation, when someone proudly showed me what he'd photographed earlier: a male pheasant. They were waiting for it to reappear.

On my morning walks over the next few weeks, I regularly encountered groups of equally intent wildlife photographers. The lake was a favourite spot, with cameras and their owners strung along its bank at intervals, waiting patiently for a bird to appear. A family of whooping swans was another favourite, as they stopped by on the migration to northern Russia.

Of course, I've met wildlife photographers before, but the number and dedication of these in Beijing tells me this is a very Chinese passion. *Nature and Wildlife Photography in China* is confirmation. It's a thrilling collection of photographs of mammals, birds, flowers, insects, and scenery. China photographers have much to be passionate about.

This group of photographers behind this collection formed a group "Imaging the Biodiversity Expedition" in 2009. They combine visuals with field surveys to document biodiversity inventories in different areas, including 60 nature reserves throughout China. They also pair with science team to make sure their identifications were right. This book summarizes their efforts over the past 10 years, covering more than 300,000 km and recording 8000 species.

China's wildlife reflects the country's exceptional range of habitats.There are black-necked cranes, descending past a snow-covered Himalayan peak, Tibetan asses and Przewalkski's horses roaming wild grasslands. There are elephants, monkeys and birds from steamy tropical forests, the Yangtze river dolphin, tropical corals. China spans the very cold to the hot, the very dry deserts to wet forests and all the species these areas contain. In an area roughly the same size as the continental United States, China has 15% of the world's terrestrial vertebrate species, and 12% of its plants. Wildlife photographers have so many species to capture!

The plant photographs alone capture this: we see them flowering on high mountains during the few weeks when the snow melts, orchids in the rainforests, flowers of dry deserts, flowers floating on lakes. Each is jewel, adding splashes of colour to the landscape.

The book's photographers stalk their animals with large telephoto lenses, of course — how else would one get a photo of a pair of snow leopards? But this book spans so many other techniques. The close-up photographs of insects are of a world with creatures beyond the imagination of those who invent science fiction. Surely the fly, Eudorylas, is a mechanised monster from another planet. The mantis, Ceratomantis, surely gives the small insects on which it feeds nightmares.

I'm pleased too to see a collection of photographs from camera traps. These are not artistic triumphs as photographs. But, they are scientifically significant — and represent a major technical advance. A decade or so ago, digital cameras were too slow to turn on, when an animal passed them. Film cameras worked but one had to check them frequently. Now, digital cameras can wait patiently for months, snapping thousands of photos. What they see are animals that we know very poorly indeed, species that are very shy and hard to detect any other way. We are learning so much about what's out there, hiding from us.

Throughout, this book fills me with a sense of wonder. For example, there are more than ten pages of different kinds of pheasants, each more beautiful than the next. I want to see that one! No, that one! China is the pheasant capital of the world with 35 species. Their homes in the mountains of the south and west are a fitting backdrop to their vivid colours. There are the herds of Tibetan antelopes, or the family of golden snub-nosed monkeys that would be fascinating to watch for hours were one lucky to come across them. Or the brilliant flowers in their alpine meadows that would fascinate on a mountain hike.

All this wonder comes with the price of worrying whether our children will be able to experience — and photograph — nature in the way we have done. Some the species are very rare indeed — the giant panda is the most famous example. It was once widespread in China but now lives in a few mountain ranges mostly in Sichuan. Worse, it's population is fragmented into isolated groups, some likely too small to be viable in the long term. Other species live in the tropical forests in Yunnan where extensive areas are being cleared for rubber plantations. The Yangtze river dolphin may now be extinct, a victim of massive industrial activity along the river's banks and the almost three-quarters of a billion people living in its watershed.

Certainly, there is some good news. China is protecting more of its land, setting aside large for nature protection. Over the last few years, I've been fortunate to visit many such places, from the wetland in Zha Long in the north to the remaining forests on the tropical island of Hainan and west to the Sanjiangyuan National Nature Reserve the Tibetan plateau in Qinghai, and Jiuzhaigou Nature Reserve and other protected places in Sichuan. Many millions of people now visit these areas. The facilities for tourists — armed with their cameras to photograph wildlife or just smartphones — are impressive indeed. Most of these areas are in the west and southwest, as indeed are the photographs in this book. There are fewer people there of course, but there are also more species.

Recent studies have shown that the species that are uniquely Chinese — those found nowhere else on Earth — concentrate in these western regions. Protecting them is vital to protecting China's biodiversity. Of course, this is where the giant panda lives and it has been a priority for the Chinese authorities.

This makes me ask: is protecting the panda good for protecting other species? Working with my Chinese colleague, Binbin Li, we've concluded that it is — a lot of China's other unique species live where the pandas live, under its protective umbrella, as it were. It's not a perfect solution – there are a lot of species that live in other places, too, and they need protection. We love pandas, but we also love the many other species that this book photographs.

And that's the point of it. China's wildlife is beautiful and fascinating and scientifically interesting and its natural areas are unique places that fascinate Chinese and international visitors alike. This book shows what we have to lose if we are not careful to protect nature.

So, when I met the group waiting for the pheasant, I was delighted. I wish I could have told them how thrilled I was to see their pleasure in such a lovely bird. With that enthusiasm comes the commitment we can all share to save China's exceptional wildlife for our children — and for thousands of more years in this country's long cultural experience.

Doris Duke Professor of Conservation Ecology

地理分区
Ground partition

东北区

东北区在中国的最北端，气候寒冷而湿润。植被从松辽平原和三江平原的草原、沼泽缓慢地过渡到大小兴安岭和长白山的森林，地势平缓，景观开阔，是驯鹿、狍、榛鸡、松鸡、东北小鲵、林蛙等典型北方物种在中国的主要分布区。在大兴安岭北部，驼鹿、雪兔和森林旅鼠等寒带典型种类也有分布，广阔的森林和众多的沼泽为有蹄类、鹤类、鹭类、雁鸭类、松鸡和雉类提供了良好的栖息地，不少在南方越冬的鸟类，夏季会迁到本区繁殖。不过，寒冷的气候限制了两栖类和爬行类物种往本区扩展，因而种类相对较少。在最近的 100 年内，森林采伐、湿地排干和农田的开垦大面积地改变了原生植被景观，已经严重威胁到东北虎、远东豹、丹顶鹤等物种的生存。

Northeast China

Located at the northern tip of China, the region features a frigid and moist climate.From east to west, steppes and marshes in Songliao and Sanjiang Plains gradually give way to forests in Greater and Lesser Khingan Ranges. Northeast China is the country's main habitat of reindeer, roe deer, (hazel) grouse, Korean salamander, wood frog and other typical northern species. The north of Greater Khingan Rangesis home to typical frigid-zone species including moose, mountain hare and wood lemming. The expansive forests and numerous marshes have created a wonderful kingdom of hoofed animals, crane, egret, wild goose, duck, grouse and pheasant. Birds overwintering in the south will migrate to the region in the summer. Daunted by the frigid climate, amphibians and reptiles here are limited in type. Over the past century, the landscape of Northeast China has been dramatically reshaped by deforestation, reclamation, and draining of wetlands. Human factors as such haveposed severe threats to the survival of Siberian tiger, Amur leopard and red-crowned crane.

< 郭亮 Guo Liang
赤狐
Vulpes Vulpes

胜山国家级自然保护区，位于黑龙江小兴安岭山地西北端，为低山丘陵加河谷湿地的地貌类型。这里是中俄边境区域内唯一森林类型保护区，也是红松林生长的北纬最上限。秋季的清晨，雨过天晴，越过对面的山岭的阳光，被层层叠叠的树冠梳理成琴弦，任由河谷飘散的雾气抚摸着、弹拨着，和着欢快的鸟鸣与潺潺的水声，奏响一曲天籁之音。

Located at the northwestern edge of Lesser Khingan Mountains, Shengshan National Nature Reserve is made up of hills, river valley and wetland. It is the only forest reserve along the China-Russia border, and the north point where Korean pine woods can grow. On an early autumn morning after the rain, the sun climbs above the yonder hills and emanates its rays through layers of tree-crowns. The mist hanging over the valley caresses the string-like rays, playing a celestial concerto with merry twitters and gurgling streams.

< 崔林 Cui Lin
东北区 – 中国黑龙江省胜山
Sheng Mountains, HeiLongjiang

华北区

华北区位于古北界的南部，南边以秦岭淮河为界与东洋界相邻，是南北动物混杂的区域，动物区系内兼有东北区森林和蒙新区草原荒漠的物种，以及喜湿的南方物种。本区特有的种类较少，只有无蹼壁虎、山噪鹛、褐马鸡、麝鼹等。区内北部山地和西部黄土高原地势较高，东部的黄淮平原地势较低，属暖温带气候，受季风影响，本区夏热冬寒，四季分明。本区人类活动和农业开发的历史极为悠久，原始森林已被栎和松的次生林所替代，原始草原已被次生草、灌丛和农田所替代，适合原始森林草原环境的大型动物，包括马鹿、梅花鹿、麋鹿、虎、黑熊等，都已在本区消失或近乎消失，数量较多的则是体形较小，或适应了农耕环境的种类，如仓鼠、田鼠、鼢鼠、乌鸦、喜鹊等。本区的沿海地带，是许多鸟类的迁徙路线，对鸟类有特殊的意义。

North China

Located in the south of the Palaearctic realm, North China shares its southern border - Qinling Mountains and Huaihe River - with the Oriental realm. It is where northern species - typified by forest creatures from Northeast China and desert/grassland animals from Inner Mongolia and Xinjiang - intermix with moisture-loving southern ones. Peking gecko, plain laughingthrush, brown-eared pheasant and desman are among the few species endemic to the region. Topographically, it is high in the north (mountainous areas) and west (Loess Plateau) and low in the east (Huanghuai Plain). Featuring a warm-temperate climate, the monsoon-influenced region has four distinct seasons. Agriculture and other human activities go back a long way in North China. The passage of years has seen primary forests give place to secondary woods of oaks and pines, and primary steppes to coppices, farmlands and meadows. Large animals which originally lived here, like red deer, sika deer, David's deer, tiger and black bear, have fully or substantially vanished; only smaller ones or those accustomed to the agri-environment, like hamster, field mouse, zokor, crow and magpie, remain in large quantities. As a part of the migration routes, coastal areas of North China are vital to birds.

< 郭亮 Guo Liang
西伯利亚狍
Capreolus Pygargus

箭扣长城位于北京市怀柔山区，沿陡峭山势起伏蜿蜒，险峻壮美。夏日雨后的黎明，湿润的空气凝结成低空的云雾漫过山脊，丝般缥缈的云海与城墙上繁茂的植物，融合成一幅梦幻的中国水墨长卷。

Beijing Jiankou Great Wall winds up and down the precipitous ridges in Huairou District. At dawn after a summer rain, the low-slung mists rimming the peaks blend with the luxuriant plants peeping above the wall into as dreamy a view as an extensive scroll of Chinese ink-wash painting.

^ 崔林 Cui Lin
中国北京市箭扣长城
Beijing Jiankou Great Wall

蒙新区

蒙新区在中国的北方，从内蒙古大兴安岭西侧一直横亘到新疆的天山——阿尔泰山。在始新世（5500 万年前—3370 万年前），本区气候温暖，雨水充沛，非常适合植物和动物的生长。随着青藏高原的隆起，到中新世（2380 万年前—532 万年前）中期，这里气候变得干旱，主要植被类型也演变为荒漠和草原，动物以适应荒漠和草原的北方型物种为主，如普氏野马、野骆驼、普氏原羚、蒙原羚、鹅喉羚、沙鼠、跳鼠、鼠兔、百灵、沙鸡、沙蜥、麻蜥等。适合草原和荒漠的啮齿类、有蹄类、爬行类和依赖啮齿动物的猛禽如金雕、大鵟等较为繁盛。间断分布在区内的水域、山地森林和草原中的有马鹿、盘羊、猞猁、雪豹、大鸨等物种。区内野骆驼为目前世界上仅存的野生种，四爪陆龟、伊犁鼠兔、新疆北鲵等物种的数量都比较少。

Inner Mongolia-Xinjiang

Inner Mongolia-Xinjiang extends from the west of Greater Khingan Ranges to Tianshan-Altai Mountains in northern China. In the Eocene epoch, it enjoyed a warm climate and ample rainfall, and served as a suitablehabitat for animals and plants. However, with the bulging of Qinghai-Tibet Plateau, the region had turned arid by the mid-Miocene epoch. Deserts and steppes became the main landforms. Drought-tolerant northern animals dominate the region, such as Przewalski's horse, wild camel, Przewalski's gazelle, Mongolian gazelle, goitred gazelle, gerbil, jerboa, pika, lark, sandgrouse, sand lizard and racerunner lizard. Other prosperous species include rodents, hoofed animals, reptiles, and rodent-eating raptors like golden eagle and upland buzzard. Waters, mountain forests and steppesare sprinkled with red deer, argali, lynx, snow leopard and great bustard. Wild camel in Inner Mongolia-Xinjiang turns out to be its only escape in the world. The region is also home to endangered species like Afghan tortoise, Ili pika and Central Asian salamander.

< 郭亮 Guo Liang
黑尾地鸦
Podoces Hendersoni

2016 年，IBE 在阿拉善盟左旗地区进行为期 10 天的影像调查。为应对沙漠的侵袭，阿拉善 SEE 基金会于十几年前开始通过种植梭梭树等方式进行治沙止漠。

In 2016, IBE conducted a ten-day image survey in Alxa Left Banner. To tame desertification, a local foundation has organized the planting of Haloxylon ammodendron for a dozen years.

< 郭亮 Guo Liang
中国内蒙古自治区阿拉善盟
Alxa, Nei Monggol

青藏区

青藏区包括青海、西藏和四川西部，由东面的横断山脉的北端，南面的喜马拉雅山脉，北面的昆仑、阿尔金和祁连各山脉环绕，平均海拔在4500米以上，是世界上最大的高原。区内分布最广的是高山草甸、高山草原和高寒荒漠，为典型的高原气候，长冬而无夏，植物生长期短促。在东南边缘则有森林和草原交错，自然条件较好。动物种类贫乏，主要由适应高原环境的特殊北方种类组成，包括藏羚羊、野牦牛、藏野驴、雪豹、兔狲、藏原羚、鼠兔、黑颈鹤、雪鸡等，爬行类和两栖类种类贫乏，仅有沙蜥、温泉蛇、西藏蟾蜍等少数种类，区内尚有大片的无人区以及羌塘、三江源、可可西里、阿尔金山、色林错、珠穆朗玛峰、工布等面积较大的自然保护区。在全球气候正在快速变化期间，青藏区所受的影响尤为显著。

Qinghai-Tibet

Encompassing Qinghai, Tibet and western Sichuan, the world's highest plateau–with an average elevation of over 4,500m–is surrounded by Hengduan Mountains in the east, Himalayan Mountains in the south, and Kunlun, Altyn-Tagh and Qilian Mountains in the north. Alpine meadows, steppes and deserts are the main landforms of the region. Featuring a typical plateau climate, it has long winters and no summer. The vegetation periods here are short. The environmental conditions of its southeast rim are relatively agreeable, with alternating forests and steppes. Plateau animals from China's north, such as Tibetan antelope, wild yak, Tibetan wild ass, snow leopard, Pallas's cat, Tibetan gazelle, pika, black-necked crane and snow cock, are preponderant among the limited species in Qinghai-Tibet. Only a few types of amphibians and reptiles can be found here, including sand lizard, hot-spring snake and Tibetan toad. The region still has some expansive uninhabited areas and China's largest natural reserves (Changtang, Sanjiangyuan, Hoh Xil, Altun Mountains, Siling Co, Mount Everest and Kongpo), yet it bears the brunt of the ongoing global climate change.

< 郭亮 Guo Liang
黑颈鹤
Grus Nigricollis

南迦巴瓦峰地区是众多雪山聚集的地方，地跨热带和寒带。在世界最后的秘境——雅鲁藏布江大峡谷的影像调查中，队员李成的红外触发相机，拍摄到一种猕猴，随后研究人员确定其为新种——白颊猕猴，所拍摄的图片及视频成为重要证据之一。

Spanning the frigid and torrid zones, the Namcha Barwa region presents concentrations of snowy peaks. During an image survey at Yarlung Zangbo Grand Canyon, one of the earth's last secret places, the infrared-triggered camera of Li Cheng, an IBE member, captured a macaque which was later identified, largely based on the images he got, as a new species, Macaca leucogenys.

< 郭亮 Guo Liang
中国西藏自治区南迦巴瓦峰地区
Namcha Barwa, Tibet

西南区

西南区是青藏高原向东和向南下降中的高山峡谷带，北起青海和甘肃南缘，南至云南北部横断山脉。本区兼具南北两方的动物区系特征，高山的动物多为北方种类，低山河谷的动物则为南方种类。区内数条南北向的山脉与河流并列，峡谷深邃，海拔高差大，植被的垂直分布明显，低山热带雨林、针阔混交林、高山针叶林、草甸和裸岩并存，为动物提供了多样化的生境，食虫类、雉类、画眉等种类非常丰富，还有大熊猫、金丝猴、羚牛、赤斑羚、藏马鸡、绿尾虹雉、高原蝮、山溪鲵等特有物种。在更新世（181 万年前—1 万年前）冰期与间冰期横扫北方时，本区并未被大面积冰盖覆盖，自然景观与现代类似，为北方动物提供了避难地，加上平行峡谷和高海拔的山峰的隔离作用，本区有利于动物的保存和分化，是世界上生物多样性最丰富的区域之一。

Southwest China

A downward extension of southeastern Qinghai-Tibet Plateau, Southwest China extends from the southern rim of Qinghai and Gansuto Hengduan Mountains in northern Yunnan. The region is featured by rivers running north to south in deep parallel valleys separated by high mountains. Vegetation here shows an obvious vertical distribution. Tropical rainforests, mixed coniferous forests, alpine coniferous forests, meadows and bare rocks provide diversified habitats at different elevations. Animals in the region originated from both China's north and south. While northern species vein the mountains, southern creatures rule the hills and valleys. Not only does Southwest China abound in insectivores, pheasant and thrush, but it also nourishes endemic species like panda, snub-nosed monkey, takin, red goral, Tibetan eared pheasant, Chinese monal, Tibetan pit viper and stream salamander. In the Pleistocene epoch, the glacial and interglacial periods which engulfed China's north seemed to have lost their bite here. Rather than being blanketed by huge ice sheets, the region maintained its landscape till the present time and turned out to be a shelter for northern species. The isolating mountains and valleys further redounded to the conservation and evolution of animals, making Southwest China one of the most ecologically diverse regions in the world.

< 郭亮 Guo Liang
赤斑羚
Naemorhedus Baileyi

梅里雪山，藏族人称之念青卡瓦格博，位于世界著名的横断山脉的西南部，是云南的最高峰，也是藏族人心中著名的神山。

Mainri Snow Mountain in the southwest of Hengduan Mountains marks the highest point in Yunnan. It is revered as a sacred mountain by Tibetans.

< 郭亮 Guo Liang
中国云南省梅里雪山
Mainri Snow Mountain, Yunnan

^ 徐健 Xu Jian

白鳖豚

白鳖豚是中国特有的淡水豚类，许多人都不曾见过它的真容。画面中的白鳖豚“淇淇”，是中国唯一人工饲养并成活的白鳖豚。它是1980年1月11日由渔民在洞庭湖捕获的，在中国科学院武汉水生生物研究所的水池里生活了22年。

Lipotes Vexillifer

The baiji (*Lipotes Vexillifer*) was a species of freshwater dolphin endemic to China; most people never met it. "Qiqi" in this picture was the only baiji that was artificially fed in China. She was captured by fishermen in Dongting Lake on Jan. 11, 1980 and spent 22 years in the pool of the CAS Institute of Hydrobiology in Wuhan, Hubei.

华中区

华中区包括西部的四川盆地、贵州高原及东部的长江流域。该区的西半部主要是山地和高原，东半部是平原和丘陵地带。整个区内有大面积的喀斯特地貌，气候温和，雨水充沛，适合植物生长。长江和洞庭湖、洪泽湖、鄱阳湖为大型鱼类、爬行类和鸟类提供了良好的栖息地。本区动物种类丰富，物种组成上表现出南北混杂和过渡的特征，东洋界的成分更多，同华南区相比，热带的物种少。白鱀豚、江豚、黔金丝猴、獐、黑麂、麋鹿、白颈长尾雉、丹顶鹤、朱鹮、扬子鳄、大鲵、中华鲟、胭脂鱼等都生活在本区。区内农业开发的历史甚为悠久，原始的亚热带森林大多为次生林、草地和农田所替代，林栖的大型动物已显著减少。近几十年中，长江及其支流上大规模的水电建设和航运发展，已经导致白鱀豚功能性灭绝，江豚、中华鲟等的生存也岌岌可危。

Central China

Central China extends from Sichuan Basin and Guizhou Plateau to the Yangtze River Basin. The western half of the region is dominated by mountains and plateaus, and the eastern half by plains and hills. Expansive karst areas can be found here. Enjoying a mild climate and ample rainfall, Central China is a suitable habitat for plants. The Yangtze River, Dongting Lake, Hongze Lake and Boyang Lake have created a wonderful kingdom of large fish, reptiles and birds. Northern and southern animals intermix and evolve in the ecologically diverse region. Species from the Oriental realm are flourishing, while tropical creatures are less preponderant here than in South China. The region is home to Chinese river dolphin, cowfish, gray snub-nosed monkey, river deer, tufted deer, David's deer, Elliot's pheasant, red-crowned crane, crested ibis, Yangtze alligator, giant salamander, Chinese sturgeon and common sucker. Agriculture goes back a long way in Central China. The passage of years has seen primary subtropical forests giving place to secondary woods, meadows and farmlands. As a result, the quantity of large arboreal animals has fallen sharply. To make matters worse, large-scale hydropower construction and shipping development over the past few decades have triggered functional extinction of Chinese river dolphin and endangered cowfish and Chinese sturgeon in the Yangtze River and tributaries.

窄脊新鼠海豚

窄脊新鼠海豚又名长江江豚。傍晚时分，长江天鹅洲的江面上，一只江豚从江水中探出身子。江豚是一种淡水豚类，它们也是中国长江生态系统旗舰物种和指标性物种，更是继白鱀豚灭绝以后亟待保护的珍稀物种。

Neophocaena Asiaeorientalis

The narrow-ridged finless porpoise (*Neophocaena Asiaeorientalis*), or Yangtze finless porpoise, is a new kind of freshwater dolphin. It is a flagship and indicator species of the Yangtze River ecosystem, and is in urgent need of protection after the extinction of baiji. In the Tian-e-zhou Oxbow of the Yangtze River, a finless porpoise leaned out of the water to embrace the evening glow.

程斌 Cheng Bin

华南区

华南区包括云南与广东、广西的大部分，福建的东南沿海以及台湾、海南和南海诸岛。区内地形多样，以低山丘陵为主，还有相当面积的喀斯特地貌。华南区处于热带和亚热带，热带鸟类繁盛，爬行类和两栖类的种类都超过其他分区。区内有亚洲象、蜂猴、白头叶猴、长臂猿、坡鹿、椰子狸、斑林狸、食蟹獴、犀鸟、原鸡、绿孔雀、巨蜥、鳄蜥等动物。原生植被主要为热带雨林和季雨林，炎热多雨的气候非常利于植物生长，森林层次结构复杂，为树栖动物提供了良好的食物和隐蔽条件，灵长类、啮齿类、食果的翼手类、食肉类中的灵猫科和鼬科，以及两栖的树蛙类动物的种类都很多。本区海岸线长，海域面积大，热带海岛众多，是我国珊瑚礁和红树林，以及儒艮、中华白海豚、棱皮龟、玳瑁、鲣鸟等物种的主要分布区。

South China

South China encompasses most of Yunnan, Guangdong and Guangxi, the southeastern coast of Fujian as well as Taiwan, Hainan and the South China Sea islands. Sizable karst areas can be found in the topographically-diverse, hill-dominated region. Featuring a (sub) tropical climate, it abounds in tropical birds and accommodates more types of amphibians and reptiles than other regions. South China is home to Asian elephant, slow loris, white-headed langur, gibbon, Eld's deer, palm civet, spotted linsang, crab-eating mongoose, hornbill, jungle fowl, green peafowl, monitor lizard and Chinese crocodile lizard. Native vegetation here mainly includes tropical rainforests and monsoon forests. Its hot and rainy days are ideal for plants, and its structurally complex forests provide food and shelter for arboreal animals. As a result, there is no shortage of primates, rodents, bats, civets, weasels and tree frogs in the region. With a long coastline, large sea areas and countless tropical islands, it boasts the largest number of coral reefs, mangrove forests, dugong, Chinese white dolphin, leatherback, hawksbill and booby in China.

< 郭亮 Guo Liang
白头叶猴
Trachypithecus Policephalus

中国南海海水清澈，透明度极佳，即使是在水深超过 10 米的地方，依然大面积分布着种类丰富的造礁石珊瑚、壮实鹿角珊瑚、芽状鹿角珊瑚等，它们在海底沟壑的顶端，抢占着有利的位置。

Through the pellucid surface of South China Sea, you will see a variety of hermatypic corals, husky staghorn corals and bud branch staghorn corals. Cloaking the top of seamounts, they extend their reaches straight down to over ten meters below the sea level.

∨ 吴立新 Wu Lixin
中国南沙群岛
Nansha Islands

哺乳动物
Mammals

向定乾 Xiang Dingqian

大熊猫

初春飞雪的清晨，陕西长青自然保护区内的一只大熊猫，为了寻找伴侣，匆忙地蹚过酉水河。由于长期跟踪和研究这只野生大熊猫，护林员与它已经适应了彼此的存在，才有机会如此近距离进行拍摄。多年默默无闻的付出，在这一瞬间值回所有。

Ailuropoda Melanoleuca

In Shaanxi Changqing Nature Reserve, a giant panda (*Ailuropoda Melanoleuca*) hurried across the Youshui River, looking for its mate amid the snowflakes of early spring. Years of tracking and study had made the ranger a comfortable company to this wide panda and made this close-up filming possible. The captured moment definitely repaid all the obscure hard works of the ranger.

< 徐廷程 Xu Tingcheng

小熊猫川西亚种

为了取回高海拔地带的红外相机，我们必须穿越原始森林。四川卧龙的秋天景色宜人，粗壮的针叶林中生长着挂满果实的花楸树，它白色的小果实是小熊猫的最爱。茂密的丛林中，贪吃的小家伙爬到花楸树上吃果实。

Ailurus Fulgens Styani

To fetch the infrared camera from a high altitude, we had to travel through a virgin forest. The autumnal beauty of Wolong baffled any descriptive words. Among the thick coniferous woods, we found Sorbus Pohuashanensis trees laden with white fruits - the most favorite food for red pandas (*Ailurus Fulgens Styani*). Actually, these greedy little ones were only spotted when climbing up the trees for fruits!

> 刘思远 Liu Siyuan

棕熊西藏亚种

多年的青藏高原探险旅行经历证明，西藏羌塘是遇见棕熊概率最高的区域。一个傍晚，在羌塘腹地多格错仁强错西侧，一个小时之内就出现了 3 头棕熊，它们有冬眠的习性，此时正在寻找尽可能多的食物，为漫长的冬眠储备能量。

Ursus Arctos Pruinosus

As proven by our years-long expedition through the Qinghai-Tibet Plateau, brown bears (*Ursus Arctos Pruinosus*) were the most frequently encountered in Qiangtang. On one particular evening, three brown bears appeared within one hour on the west of Doké Tsoring Jangtso, the hinterland of Qiangtang. Those hibernators had to find as much food as they could to conserve energy for the long winter.

∧ > 董磊 Dong Lei

西黑冠长臂猿

生活在云南景东无量山国家级自然保护区的西黑冠长臂猿，是中国现存 6 种长臂猿中的一种。中国境内分布的 6 种长臂猿，两种已经多年不见。唯一数量较多的只有西黑冠长臂猿，但总数比大熊猫还要少。西黑冠长臂猿的雌猿与小猿是金黄色的，雄猿则全身乌黑。它们的主食是各种果实和嫩叶，不同的季节采摘不同的果实食用。

Nomascus Concolor

Inhabiting Mt. Wuliang, a national nature reserve in Jingdong County, Yunnan Province, *Nomascus Concolor* is one of the six gibbon species that can be found in China; in fact, two of its cousins have not been seen for many years. Though being the only gibbon genus that exists in relatively large quantity, *Nomascus Concolor* is even rarer than pandas. The species exhibits dichromatism where the female and baby are golden colored while the male is almost completely black. The gibbon feeds on fruits and young leaves which vary somewhat by seasons.

东白眉长臂猿

东白眉长臂猿是中国一级保护动物，最濒危的野生物种之一，目前十分罕见。在云南的森林里，清晨时分，它们会通过鸣叫来宣示自己的地盘。这种与人类亲缘关系最亲密的野生动物所能表现出来的东西，让人对大自然造物感动得无以复加。

Hoolock Leuconedys

Hoolock Leuconedys, one of the most endangered wide animals, has become a rarity in China and is placed under the first-grade State protection. Every morning when the first rays emblaze the forests of Yunnan, gibbons chant loudly to declare their territories. We are amazed to the utmost by this wonderful creature of the nature which bears the closest relationship with humans.

滇金丝猴

丽江老君山，连续的阴雨结束后放晴的当天傍晚，忽然，附近的冷杉林里传来“吱—嘎”声，一群 200 多只的滇金丝猴在寻找夜栖的大树。山谷里最后一丝阳光勾勒出一只滇金丝猴俊俏的身影。滇金丝猴长着一张最像人的脸庞，白里透红，还有一张美丽“性感”的红唇。

Rhinopithecus Bieti

In the Mt. Laojun of Lijiang, the sky finally cleared up after a long raining spell. Suddenly, a squeaking sound broke the evening murmuring of the fir forest; a flock of 200 plus black snub-nosed monkeys (*Rhinopithecus Bieti*) were searching for their night habitats among the tall trees. A pleasing figure was profiled against the very last ray that filled the valley. These monkeys' faces, dotted with white rosy cheeks and beautiful sexy lips, are startlingly human-like.

>程斌 Cheng Bin

川金丝猴

这种猴科仰鼻猴属中美丽的猴子，中文称之为金丝猴，因为它们的体毛为金色的丝状，最早在四川地区被发现，也称为川金丝猴。另外几种同科物种的毛发均为暗色。它们群栖在高山密林之中，以植物性食物为主。

Rhinopithecus Roxellana

Rhinopithecus Roxellana, a beauty from the rhinopithecus family, derives its Chinese name (literally, "Sichuan golden hair monkey") from its golden hair and "Sichuan" where it was first discovered. Those monkeys live in groups among the dense forests and highest peaks and feed on plants. They look slightly different from other cousins of the *rhinopithecus* family whose hairs are dark colored.

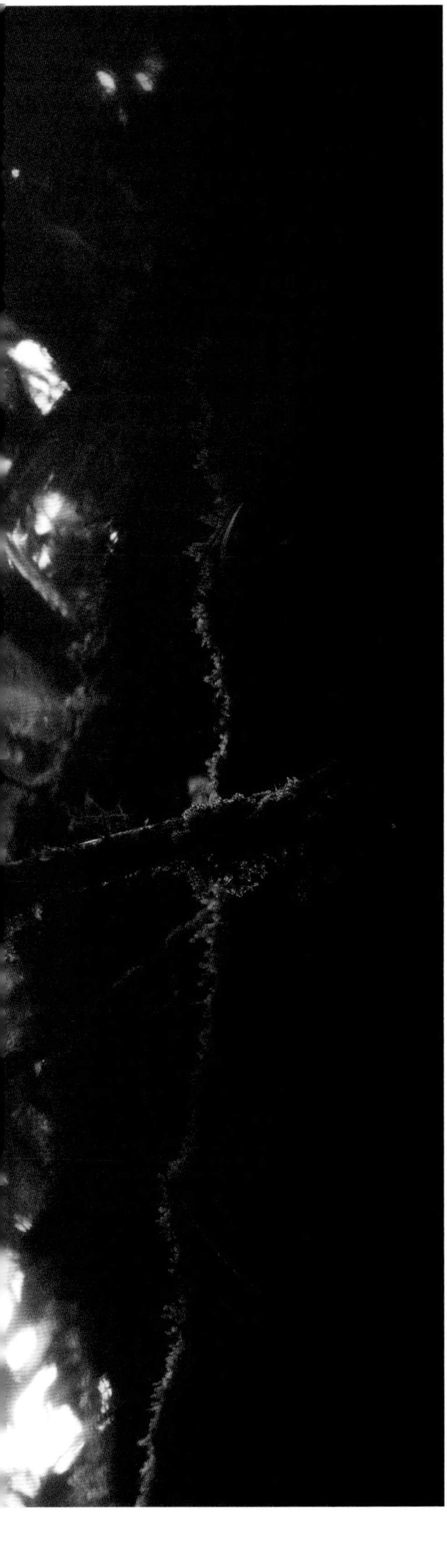

< 左凌仁 Zuo Lingren

怒江金丝猴（别称缅甸金丝猴）

云南泸水片马地区的原始森林里，一只带领猴群移动的雄性缅甸金丝猴发现了人的存在，警惕地观察着动向。真是“众里寻他千百度”，我们千辛万苦都没能找到缅甸金丝猴的踪影，没想到，在我们已经放弃寻找准备下山的时候，一次“蓦然回首”，它们却突然出现了。

Rhinopithecus Strykeri

In the virgin forests of Pianma Town, Lushui County, we came across a troop of monkeys led by a vigilant male snub-nosed monkey (*Rhinopithecus Strykeri*). The encounter could not have been more timely since we were about to return home empty-handed after a tiresome search for the very species.

^ 张兴伟 Zhang Xingwei

印支灰叶猴

云南景东无量山国家级自然保护区的印支灰叶猴。

Trachypithecus Crepusculus

Trachypithecus Crepusculus in Mt. Wuliang National Nature Reserve, Yunnan Province.

^ 吴秀山 Wu Xiushan

菲氏叶猴

云南高黎贡山国家级自然保护区的菲氏叶猴。

Trachypithecus Phayrei

Trachypithecus Phayrei in Mt. Gaoligong Nature Reserve, Yunnan Province.

^ 郭亮 Guo Liang 白头叶猴 *Trachypithecus Leucocephalus*

^ > 郭亮 Guo Liang

白头叶猴

白头叶猴是生活在广西喀斯特地貌的中国特有物种，也是世界上第一个由中国人命名的灵长类。它们以绝壁上的石窝儿、小台地为家，以植物的叶和茎为主要食物。猴王为守护领地，会从近百米落差的绝壁上飞跃而下，赶走入侵者。

Trachypithecus Leucocephalus

White-headed langur (*Trachypithecus Leucocephalus*) is endemic to China where it occupies karst cliffs in Guangxi Province. It is also the first primate species that named by Chinese. The langurs live in caves and platforms and feed mainly on leaves and stems. To affend its own territory, the langur king may leap off the cliff with a 100m sheer drop to chase the invaders away.

< 左凌仁 Zuo Lingren

喜山长尾叶猴

西藏樟木地区的树林中，一只喜山长尾叶猴惬意地坐在树枝上，打量着周边活动的同类。这是我们多年以来第一次遇到“不认生”的野生灵长类动物。

Semnopithecus Schistaceus

Contentedly, a gray langur (*Semnopithecus Schistaceus*) sat on the limb of a tree, looking over to its friends playing in the woods of dram, Tibet. For the first time in years, we finally came across a wild primate who was comfortable with strangers.

白颊猕猴

在西藏自治区墨脱县与世隔绝、浓密幽暗的森林里，红外线触发照相机拍摄到许多很难直接目击的大型哺乳动物，它们的种类、数量都颇为可观，其间，最引人注目的就是新发现的灵长类物种——白颊猕猴。

Macaca Leucogenys

Through the dense and dim woods, which seemed a cloistered world in Motuo County in southeastern Tibet, the camera traps captured the images of many large mammals which were hard to be observed directly. Those mammals existed in considerable quantities and varieties, where the newly discovered white-cheeked macaques (*Macaca Leucogenys*) really caught our eyes.

^ 李成 Li Cheng

郭亮 Guo Liang

亚洲象

亚洲象是中国陆地上最大的哺乳动物，现仅生活在云南西双版纳地区。十几年前，一小群亚洲象向北渡过澜沧江，到达普洱思茅地区。此后便经常在当地村庄、田地和山林间游走，水稻田是它们经常光顾的地方。

Elephas Maximus

Asian elephant (*Elephas Maximus*) is the largest land mammal in China. It lives only in Xishuangbanna, Yunnah Province. A dozen years ago, a small group of Asian elephants travelled north across Lancang River to Simao District in Pu'er Prefecture. Since then, they have shuttled between local villages, farmlands, and mountain forests. Paddy fields seem one of their favorite.

^ 徐健 Xu Jian

普氏野马

在新疆准噶尔盆地的卡拉麦里保护区内，两匹珍稀的普氏野马在静静地觅食。这里曾是它们祖先奔腾的原野，它们曾经是中国西部和蒙古荒漠草原上的典型物种。野生普氏野马已经于 20 世纪 60 年代灭绝。

Equus Przewalskii

In Kalamaili Ungulate Protected Area, Junggar Basin, Xinjiang, two Przewalski's horses (*Equus Przewalskii*) paced quietly about for food. This was exactly where their ancestors used to gallop around. *Equus przewalskii* was once native to the steppes of western China and Mongolia, but the wild population died out in the 1960s.

亚洲野驴

即使在中国内蒙古境内，国家一级保护动物亚洲野驴也非常罕见。2005 年 11 月，我在中蒙边界线附近寻找草原狼时，无意间遇见了这群亚洲野驴。它们生活在一片类似孤岛的区域，由于没有同类加入，而周围又有草原狼环伺，生存实属不易。

Equus Hemionus

Even in the Inner Mongolia, the Asian wild ass (*Equus Hemionus*) rarely exists. It is now under first-class State protection. In Nov. 2005, we stumbled on a group of wild asses when searching for coyotes along the China-Mongolia borderline. Their existence on this isolated place became incredibly hard without the company of fellow creatures and under the threats of coyotes.

∨ 刘思远 Liu Siyuan

程斌 Cheng Bin

藏野驴 *Equus Kiang*

< 郭亮 Guo Liang

藏野驴

在青藏高原，不畏严寒而且能够长距离快速奔跑的动物，除了著名的藏羚羊，还有更具耐力的藏野驴。它们常年生活在高原极地地区，身体已经完全适应了那里的各种气候环境。在壮美的雪山圣地，它们生活得悠然自得。

Equus Kiang

Apart from Tibetan antelope, there is another animal that defies severe cold and runs incredibly fast over long distance on the Qinghai-Tibet Plateau. That is the Tibetan wild ass (*Equus Kiang*) which displays even a greater endurance of strength than the antelope. It lives in polar regions at high altitudes all year round, and has physically inured itself to the varying climates there. It leads a leisurely life on the sacred and spectacular snowy mountains.

盘羊

盘羊是家绵羊的近亲，公羊大角如盘。2014 年，IBE 在西藏山南影像调查中拍摄到一只公羊站在高高的山岗上守护着它的羊群。

Ovis Ammon

Argali (*Ovis Ammon*) is a close relative of sheep. The male has large spiraling horns. During an image survey in Shannan, Tibet in 2014, IBE captured on camera a male guarding his flock on top of a hill.

郭亮 Guo Liang

∨ 郭亮 Guo Liang

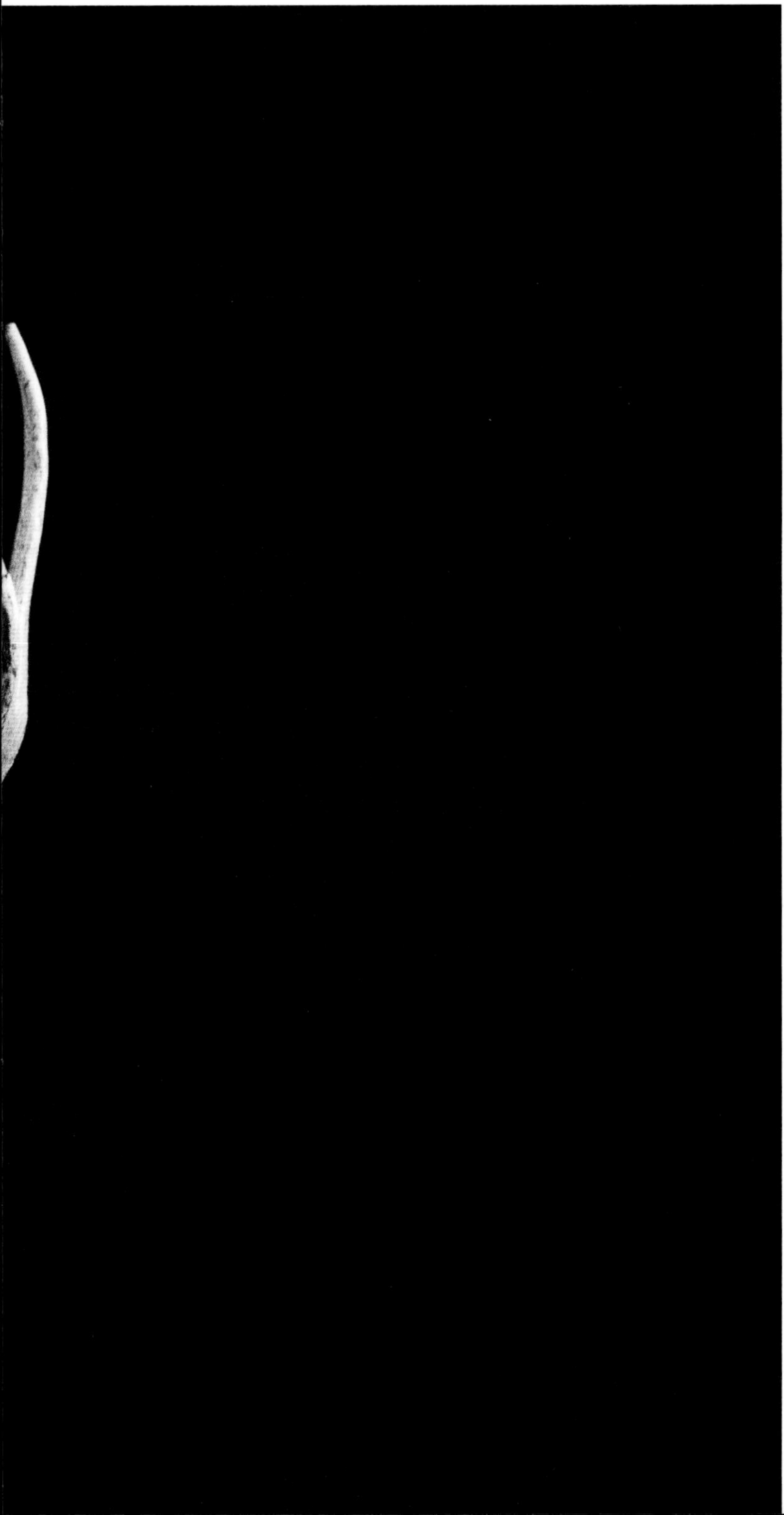

白唇鹿　*Cervus Albirostris*

白唇鹿

白唇鹿是仅中国有的一种典型的高寒动物。11 月份，公鹿不再集群，从海拔 5000 米迁徙下来，经过争斗获胜的公鹿拥有自己的鹿群和交配权。傍晚时分，鹿群在公鹿的带领下，来到沟谷里的溪流边饮水。

Cervus Albirostris

White-lipped deer (*Cervus Albirostris*) is endemic to China where it inhabits only the high and cold grasslands. The males fall out and migrate from an altitude of more than 5,000 meters every November. Males which won the duel may have their own droves and right to mate. Towards the end of an afternoon, led by a male, a drove of deer went down the valley to drink along the streams.

^ 郭亮 Guo Liang

梅花鹿台湾亚种

中国台湾，梅花鹿已经进入发情的季节。天刚亮，草原的尽头有两只年轻的公鹿对峙，不一会儿，它们便用头上的大角对撞了起来，撞击声回荡不绝。与其说这是两只公鹿的争斗，不如说是一支生命的舞蹈。

Cervus Nippon Taiouanua

In Taiwan, sika deer (*Cervus Nippon Taiouanua*) began to show signs of estrus. The day was just dawning when two young males confronted each other at the end of the grasslands. A moment afterwards, they bumped against each other with their horns, sending out a resonant sound. It was not so much a fight between two males as a dance of life.

∨ 黄一峰 Huang Yifeng

^ 董磊 Dong Lei

梅花鹿四川亚种

清晨，在四川若尔盖铁布梅花鹿自然保护区山腰，我邂逅了带着几只小鹿的母鹿，这是四川梅花鹿，直到 20 世纪 60 年代才被科学界发现，多亏偏远的地理位置和藏传佛教文化的保护，这种漂亮的动物才得以幸存。

Cervus Nippon Sichuanicus

In a fine morning, I encountered a female Sichuan sika deer (*Cervus Nippon Sichuanicus*) and her young, half way up the mountain in Tiebu Nature Reserve, Ruoergai County. The deer was not discovered until the 1960s. This gorgeous species is lucky to survive thanks to the geographical remoteness and the protection of Tibetan Buddhism.

^ 郭亮 Guo Liang

马麝

马麝是麝类中体形最大的一种。在中国西藏藏南地区，近 10 只马麝生活在一座村庄周围。清晨，它们游走在村庄与田地之间，看见人就机警地跑开，在不远处又停下，细细地打量陌生人。

Moschus Chrysogaster

The alpine musk deer (*Moschus Chrysogaster*) is the largest member of the musk deer family. In southern Tibet, nearly ten alpine musk deer lived around a village. They wandered between the village and fields in the morning, and ran warily away from any strangers before stopping to scrutinize the strangers from a little distance.

獐

獐被认为是最原始的鹿科动物，原产地在中国东部和朝鲜半岛。初春清晨，中国丹东湿地，它在刚刚长出芦苇嫩芽的开阔地带进食，发现缓慢行进的汽车，好奇地跑过来一探究竟。

Hydropotes Inermis

River deer (*Hydropotes Inermis*), the most primitive species of cervidae, is native to eastern China and the Korean Peninsula. In an early-spring morning in Dandong Wetland, a River deer was taking its breakfast in an open field which just began to push out new reed sprouts. Suddenly, it noticed the slow-moving car and came over to take a curious look.

∨ 郭亮 Guo Liang

吴秀山 Wu Xiushan

野牦牛 *Bos Mutus*

^ 吴秀山 Wu Xiushan

野牦牛

野牦牛是青藏高原上真正的霸主，也是十分危险的动物，所谓“牛脾气”用来形容它们再合适不过了。100 多只野牦牛在湖边安静地休息，突然，不知什么原因，牦牛群开始移动，扬起滚滚的沙尘，气势摄人心魄。

Bos Mutus

Wild Yak (*Bos Mutus*), a very dangerous animal, is indeed the dominator of the Qinghai-Tibet Plateau. "Bull-headedness" may be the most apposite word to describe its temper. Over 100 wild yaks settled themselves by the lake most peacefully, then somehow started to move, lifting up clouds of dusts with a show of imposing strength.

^ 郭亮 Guo Liang

四川羚牛

11 月份的一个傍晚，在四川省唐家河国家级自然保护区，我与近 20 只羚牛隔河相遇。它们在河对岸的山坡上悠闲地进食，羚牛幼崽跟着妈妈到河边饮水，我们之间只相隔 40 米。

Budorcas Taxicolortibetana

We traveled to Sichuan Tangjiahe Nature Reserve in November. When the night fell, we saw nearly 20 Sichuan takins (*Budorcas Taxicolortibetana*) eating food on the hillside across the river. Young babies followed their mothers to drink at the river. We were only 40 meters apart.

^ 向定乾 Xiang Dingqian

秦岭羚牛

清晨，我在秦岭高山草甸上寻找秦岭羚牛的踪迹，就在准备放弃时，突然看见对面的山崖上有 3 只羚牛在一起觅食。秦岭羚牛与其他种羚牛的区别在于，秦岭羚牛全身金黄通透，当地人也叫它“金毛牛”。

Budorcas Bedfordi

We were about to return empty-handed in the morning from the search for golden takins (*Budorcas Bedfordi*) across the alpine meadow of Qinling Mountains, when three takins ranging together over the cliff jumped into our view. Qinling Golden Takin is given this name for its long golden hair round the body, which distinguishes it from other species.

范毅 Fan Yi

高黎贡羚牛

初秋，清澈的湖蓝色河水穿过云南高黎贡山渐黄的森林，这天然又美丽的地方是一群羚牛的家园。

Budorcas Taxicolor

The crystal blue stream ran through the woods of Mt. Gaoligong in early autumn, a beautiful home to a crowd of takins.

藏原羚

Procapra Picticaudata

∨ 郭亮 Guo Liang

< 郭亮 Guo Liang

藏原羚

藏原羚是典型的高山寒漠动物，是青藏高原特有的物种，俗称“白屁股”。在西藏行车时，经常可以看到它们三五成群走过的身影。

Procapra Picticaudata

Procapra Picticaudata is a typical alpine animal that inhabits only the cold deserts of the Qinghai-Tibet Plateau, and has a white patch on its rump. When driving in Tibet, people can easily encounter them in small or large groups.

< 吴秀山 Wu Xiushan

普氏原羚

普氏原羚很有“故事”，它的发现、命名到保护，都足以用长篇大论来讲解。相对于其他近缘物种，它们最与众不同的是对角，有人称之为“中华对角羚”，这也是在野外最容易远距离识别它们的标志之一。

Procapra Przewalskii

Przewalski's Gazelle (*Procapra Przewalskii*) has many long stories to share about its discovery, naming and protection. One of the most important markings that distinguishing it from its cousins, even viewed from long distance in the wide, is the most curvaceous horns.

郭亮 Guo Liang

赤斑羚

赤斑羚被列为中国一级保护珍稀物种。雅鲁藏布江大峡谷的奇特地貌为赤斑羚的生存提供了保护，水流湍急的雅鲁藏布江成为赤斑羚的天然保护屏障。赤斑羚来到江边饮完水后，很快消失在巨石之后，隐遁在树林中。

Naemorhedus Baileyi

The red goral (*Naemorhedus Baileyi*) is an endangered species under the first-grade State protection in China. The turbulent river crashes its way through the Yarlung Zangbo River, forming a safeplace for red gorals to live. A red goral finished drinking at the river, not long before it disappeared behind a huge rock and melted into the woods.

∧ > 吴秀山 Wu Xiushan

藏羚

西藏羌塘的傍晚时分，一只雄性藏羚羊在追求一只雌性藏羚羊，随着母羊的移动，公羊寸步不离地追逐，母羊突然越过了公路，公羊穿越时，在日落的光影下，公路弯曲的曲线正好形成了笼罩在藏羚羊周围的光环。

Pantholops Hodgsonii

In Qiangtang, a male Tibetan antelope (*Pantholops Hodgsonii*) paid court to a female at twilight. Unexpectedly, the female crossed the road; the male followed closely and was beautifully profiled against the winding roads as if with a halo.

藏羚

阴历十六，青海可可西里自然保护区，月亮还没有完全落下，太阳已升起，日出时特有的柔和光线打在一群觅食的母藏羚羊身上，随着藏羚羊的远去，月亮渐渐消失在如梦幻般的日光里，整个过程唯美而充满诗意。

Pantholops Hodgsonii

It was the sixteenth day of the lunar calendar. In Kekexili Nature Reserve, the sun had risen above the horizon before the moon went completely down, throwing soft lights onto a group of female Tibetan antelopes which were searching for food. As the foragers went away into the distance, the moon melted into the daylights, dreamily with poetic grace.

岩羊

大雪后，进行动物调查的我们刚刚爬上山脊，发现对面山上有影子在移动，原来是一群岩羊发现了我们，母羊领头，带着群羊希望翻过雪山离开我们的视线。然而由于积雪太厚，头羊被困，没有找到很好的离开路线，群羊也因此停了下来。我们在两个山头对峙。我们拍下几张照片后便走出它们的视线，希望母羊能带领群羊脱离困境。

Pseudois Nayaur

We started the animal study after the heavy snow. Just commanding the high ridge, we noticed moving shadows on the opposite hill - a flock of bharals (*Pseudois Nayaur*) got startled at the sight of us. Led by a female bharal, the flock attempted to run across the hill to somewhere out of our sight. But the thick snow trapped the female head, without knowing where to exit, the flock stopped. I took a few pictures and left immediately till they can't see, in a hope that the female would lead the flock out of there soon.

^ 郭亮 Guo Liang

岩羊

初冬的澜沧江源，为了拍摄雪豹，我们隐藏在岩羊群周围。岩羊是雪豹的主要食物来源，看见岩羊，意味着雪豹也许就在附近。一只公岩羊发现了我们，它居高临下俯瞰着，很好奇。

Pseudois Nayaur

One day in the early winter, we concealed around a flock of bharals (*Pseudois Nayaur*) to take the picture of snow leopards in the source region of the Lancang River. Snow leopards, which feed mainly on bharals, may probably be somewhere about. Then a male bharal noticed us, curiously overlooking us from a high place.

∨ 郭亮 Guo Liang

岩羊

Pseudois Nayaur

∨ 彭建生 Peng Jiansheng

雪豹

在青海三江源君曲山谷，快成年的孩子在山顶晒太阳。我费尽精力爬上海拔 5500 米的高处准备拍摄它们一家时，雪豹妈妈绕过山谷迎面走来。

Panthera Uncia

At the Junqu Valley in Sanjiangyuan (the headwater area of the Yangtze, Yellow and Lancang Rivers), a female snow leopard (*Panthera Uncia*) basked on the mountaintop, accompanied by her two kids on the threshold of manhood. As I exhausted all my strengths ascending the height of 5,500 meters to take a group picture of them, the mother leopard was just coming towards me around the valley.

郭亮 Guo Liang

猞猁

西藏的阿里地区，虽已是晚上8点，但离天黑尚早。一只猞猁懒散地游荡在稀疏的灌丛中，它知道傍晚时分有很多高原兔会出现在这里，那将是它今晚的主菜。

Lynx Lynx

It is 8 p.m. in Ngari Prefecture, Tibet. The night is still young. A lynx is milling around the sparse bushes, waiting for tonight's pièce de résistance, woolly hares.

郭亮 Guo Liang

兔狲

青海三江源保护区的一处高山草甸上，一只成年的兔狲谨慎地从洞口爬出，为它的 4 个孩子去捕食鼠兔。它们利用一处岩石下的旱獭洞做窝，由于身体的隐蔽色很好，而且动作机警，很难被发现。

Otocolobus Manul

On an alpine meadow in Sanjiangyuan protected area, an adult Pallas's cat (*Otocolobus Manul*) crawled carefully through the mouth of its cave, looking for pikas to feed its four babies. This family lived in a marmot hole under the rock and was hard to be detected thanks to their mimic coloration and vigilant movements.

狼

四川西部马尼干戈，天色渐渐暗了下来，高原上忽然下起了小雪。在山谷中，我偶遇这头狼，它忧郁的眼神让人久久不能忘怀。

Canis Lupus

It started to snow on the plateau as the darkness fell over Maniganggo in western Sichuan. I came across this grey wolf (*Canis Lupus*) in the valley and was so much impressed by his gloomy stare.

∧ > 郭亮 Guo Liang

藏狐

在青海年保玉则附近的一处草原，经过两个多小时的不断尝试，我终于能靠近这3只小狐狸，距离它们不到10米。“它们接受了我。在我面前，它们追逐、嬉戏，甚至睡觉。我不愿再按动快门，只想和它们四目相对，共享午后暖暖的阳光。”

Vulpes Ferrilata

On the grassland near Mt. Nianboyuze, I finally approached two little Tibetan sand foxes (*Vulpes Ferrilata*), with less than 10 meters between us, after more than two hours of constant tries. "They accepted me, chasing, playing and even sleeping in front of me. I had no wish to press the shutter button, just wanted to meet their eyes, basking together in the warm afternoon sun."

< 董磊 Dong Lei

艾鼬

沿着四川省若尔盖穿越大草原的公路拍摄时，我意外发现了两只活泼可爱的艾鼬，它们在草地上扑腾嬉戏，享受美好的时光。这两只艾鼬时而穿越草地洞穴，时而腾跃在空中，身姿矫健，无比灵活，令人叹为观止。

Mustela Eversmanii

Unexpectedly, I encountered two lively steppe polecats (*Mustela Eversmanii*) as I took photos along the road across the vast prairie in Ruoergai County, Sichuan. They were having a great time running through the caves or leaping into the mid-air, nimbly and agilely to amaze me.

^ 彭建生 Peng Jiansheng

香鼬

在长江源头沱沱河边巧遇这只香鼬，它在石堆中站立起来好奇地面对镜头，非常放松。香鼬虽然体形较小，但是行动迅捷，捕猎也非常高效。很多小型动物都是它们的猎物。

Mustela Altaica

This little mountain weasel (*Mustela Altaica*), which I came across by the Tuotuo River, stood up through a pile of stones, looking curiously into the camera. Small though, the weasel runs swiftly and hunts effectively. Its games include many smaller animals.

^ 郭亮 Guo Liang

川西鼠兔

因摄食方式和行为等与兔子相像，故名鼠兔。青藏高原为现生鼠兔分布中心与演化中心。夏季的澜沧江源地区，在高山草甸的石山上，生活着川西鼠兔，它们的行动小心谨慎，时时刻刻提防着猛禽和藏狐等天敌。

Ochotona Gloveri

It is given the Chinese name "a rabbit-like mouse" for its resemblance to rabbits in foraging habits and behaviors. The Qinghai-Tibet Plateau is where pikas inhabit and evolve. In summer, Glover's pikas (*Ochotona Gloveri*) lived on the rocky hills of alpine meadow in the source region of the Lancang River. They moved cautiously and stayed always on guard against raptors, Tibetan foxes and other enemies of the nature.

^ > 郭亮 Guo Liang

喜马拉雅旱獭

喜马拉雅旱獭广布在青藏高原，是一种大型啮齿动物。8 月中旬，青海三江源治多县夏日寺附近的高山牧场已是芳草连天，鲜花遍野，这是高原最美最幸福的时光。一对喜马拉雅旱獭正在通过戏耍的方式学习生存本领，它们相互推挡、拥抱、接吻，它们“窃窃私语”，亲密无间。

Marmota Himalayana

Himalayan Marmots (*Marmota Himalayana*), a large species of rodent, is widely distributed over the Qinghai-Tibet Plateau. It was mid-August, the best days for the Plateau, when the green grass which was strewed with flowers stretched far into the sky on the alpine pasture near the Summer Temple in Zhiduo County, Sanjiangyuan. A couple of Himalayan marmots "whispered" intimately while learning the life skills through games: pushing, wrestling and kissing.

^ 秦大公 Qin Dagong

中华白海豚

中华白海豚数量稀少，有“海中熊猫”的美誉。在中国广西的三娘湾有着健康的种群。它们与当地渔民共享大风江入海口这块海域，一只年轻的白海豚在渔船附近欢快地跳跃，久久不愿离去。

Sousa Chinensis

The Chinese white dolphin (*Sousa Chinensis*) is reputed as "the panda in the sea" for its rarity. Some healthy ones live in Sanniang Bay of Guangxi, and share the mouth of the Dafeng River with the local fishermen. A young white dolphin jumped about playfully around a fishing boat, reluctant to leave for a long time.

^ 郭亮 Guo Liang

中华白海豚

Sousa Chinensis

鸟类
Birds

徐健 Xu Jian
棕尾虹雉
Lophophorus Impejanus

郭亮

棕尾虹雉

棕尾虹雉也叫九色鸟，尼泊尔国鸟，是喜马拉雅山脉特有的大型珍稀鸟类。在西藏南部的一座寺庙里拍摄时，它竟慢慢向我们走来，我们得以近距离一睹它的“雍容华贵”。

Lophophorus Impejanus

Himalayan Monal (*Lophophorus Impejanus*), also known as "a bird of nine colors", is a large and rare species endemic to the Himalayas, and the national bird of Nepal. As I took photos around a temple in southern Tibet, a Himalayan monal walked slowly towards me and granted me a close glimpse of its radiant splendor.

∧ > 郭亮 Guo Liang

绿尾虹雉

雪后，在四川卧龙国家级自然保护区山顶的草甸乱石堆中，终于见到了梦寐以求的绿尾虹雉。“它边走边取食，几乎来到了我面前，全身的羽毛绽放着金属般的光泽，让人不禁感叹自然界的神奇。”

Lophophorus Lhuysii

After the snow, I reached the stony hilltop meadow in Wolong National Nature Reserve where I came across a Chinese monal (*Lophophorus Lhuysii*), one of my fondest wish for years. "In search of food, it walked almost up to me. The golden feathers had a metallic luster. I couldn't help to praise the miracle of nature."

白尾梢虹雉

小雪后，濒临灭绝的白尾梢虹雉在云南高黎贡山漫步。蔚蓝的天空下，它的羽毛反射出难以描述的色彩。

Lophophorus Sclateri

After a light snowfall, a Sclater's monal (*Lophophorus Sclateri*), an endangered species of bird, walked leisurely on Mt. Gaoligong in Yunnan. Its feathers took on an unbelievable spectrum of colors under the blue sky.

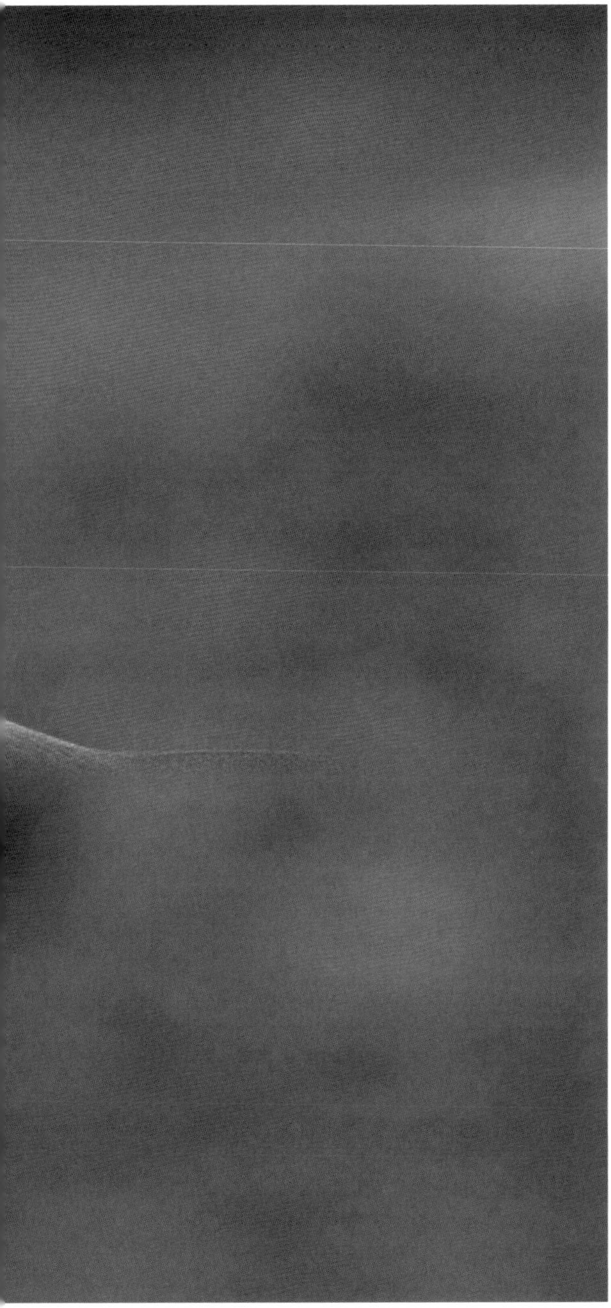

< 董磊 Dong Lei

红腹锦鸡

红腹锦鸡是中国特有鸟类，因为全身鲜艳的红黄色调而被古人称为“金鸡”。清晨，我在四川省唐家河国家级自然保护区内一条人迹罕至的林区小路上遇到了一只漂亮的雄鸟，我待在路边的排水沟里一动不动，空旷的道路在锦鸡前面有一条凸起，照片前景与背景虚化的程度很高，与锦鸡鲜艳的毛色对比产生了神秘的视觉效果。

Chrysolophus Pictus

The golden pheasant (*Chrysolophus Pictus*) is a gorgeous bird species endemic to the mountains of China, and was referred to as "golden pheasant" by the Chinese ancients for its bright reddish-yellow color. I came across a gorgeous male golden pheasant in the morning as I walked along an unfrequented path through the woods in Sichuan Tangjiahe National Nature Reserve. I stood, like a log, in the roadside gutter. The highly blurred foreground and background formed a contrast to the bright feathers, and produced a mysterious visual effect.

白腹锦鸡

清晨 6 点，我在云南梅里雪山的一个垭口静静地坐着，当地人传说中的“长尾巴野鸡”即白腹锦鸡如约而至，漫步到我身边，然后从容地进入灌丛。白腹锦鸡是青藏高原东部的特有鸟类，主要分布在中国，在森林中难得一见。

Chrysolophus Amherstiae

A Lady Amherst's pheasant (*Chrysolophus Amherstiae*) which was also referred to as "long-tail pheasant" by the local people, walked to me as if by appointment at six in the morning while I was sitting at a narrow pass of the Mainri Snowy Mountain. Leisurely, it passed me and disappeared into the bushes. *Chrysolophus Amherstiae* is native to the eastern part of the Qinghai-Tibet Plateau and is mainly distributed in China. It rarely shows up in forests.

∨ 徐健 Xu Jian

^ 郭亮 Guo Liang

藏马鸡

藏马鸡是中国特有的鸟类，分布于西藏东部、青海南部、四川西部、云南西北部，属留鸟。得益于藏传佛教的影响，藏马鸡的栖息地受到了较好的保护。

Crossoptilon Harmani

Crossoptilon Harmani is a species of resident bird endemic to China, and is confined mainly to eastern Tibet, southern Qinghai, western Sichuan and northwestern Yunnan. Its habitats are now under sound protection, which is a benefit of the Tibetan Buddhism.

^ 王斌 Wang Bin

褐马鸡

2012 年元旦，一场突如其来的大雪后，栖息在寺庙周围山上的褐马鸡纷纷下山觅食。已经习惯了和寺庙中的僧人相处的褐马鸡，对外来人没有丝毫戒备，它拖着长长的尾羽，泰然自若，闲庭信步。更难得的是，它竟然像孔雀般打开了尾羽，美得令人惊叹。

Crossoptilon Mantchuricum

After a sudden fall of snow on the first day of 2012, the brown eared pheasants (*Crossoptilon Mantchuricum*) which had sheltered in the temple went down the hill for food. Accustomed to the company of monks, the pheasants were perfectly comfortable with strangers. Wearing long tail feathers, they walked idly and composedly. More surprisingly, one of them spread the tail like a peacock, which was bewilderingly beautiful!

白马鸡

初冬的澜沧江源地区，春夏间难得一见的白马鸡开始集群。每每在清晨，它们一群一群地出现在村庄、田地周围，对过往行人、车辆没有特别的惧怕，而是欢快地嬉戏、进食。

Crossoptilon Crossoptilon

White eared pheasants (*Crossoptilon Crossoptilon*) gathered at the source region of the Lancang River as the winter came that you can rarely expect in summer and spring days. Every morning, they appeared in groups around villages and farmlands, frolicking and eating merrily, without a particular fear of passing crowds or vehicles.

蓝马鸡

四川与青海交界的麻尔柯河林场，是青海最大的森林区，麻尔柯河林场的针叶林是中国的蓝马鸡的重要栖息地。蓝马鸡的叫声引人深入林区，但是一般在茂密的森林和灌丛里，还是很难发现它们的。中国的4 种马鸡：蓝马鸡、褐马鸡、白马鸡、藏马鸡，都是中国特有种，国家级保护动物。

Crossoptilon Auritum

Makehe Forest Farm, located at a place where Sichuan and Qinghai meet, is the largest forest area in Qinghai. The coniferous woods are an important habitat for blue eared pheasants (*Crossoptilon Auritum*). People may be attracted into the thick woods and bushes by the crowing of the pheasants, but can hardly see one. The four species of eared pheasants, namely *Crossoptilon Harmani, Crossoptilon Mantchuricum, Crossoptilon Crossoptilon* and *Crossoptilon Auritum*, are all endemic to China and placed under the State protection.

血雉指名亚种

2009 年秋天的一个清晨，一群血雉在云南梅里雪山林间的小溪饮水后，来到一块小草坪上觅食。即将走出灌丛时，它们发现了隐蔽在附近的摄影师，开始四处逃窜，有的还警戒地打开尾羽伪装，其中一只跑远了之后，又回头看了一眼。

Ithaginis Cruentus

It was an autumnal morning in 2009. A group of blood pheasants (*Ithaginis Cruentus*) finished drinking along the stream in the Mainri Snowy Mountain and moved to a small lawn for food. They were just about to walk out of the bush when catching sight of the photographer who concealed somewhere nearby. Frightened, they ran in all directions; some even spread their tails as a disguise. Interestingly, one looked back when running well away.

∧ 郭亮 Guo Liang

环颈雉

环颈雉是中国雉科中分布最广的鸟。丹东湿地的初春，是雉鸡恋爱的季节。每每在湿地的路上行走，我们总是会听到雄鸟在干芦苇丛中鸣叫，它们有时又会从旁边的灌丛中突然飞起。

Phasianus Colchicus

The ring necked pheasant (*Phasianus Colchicus*) is the most widely distributed among all members of the Phasianidae family. On the wetland of Dandong, pheasants fell in love in early spring. Every time we walked along the path, we heard male pheasants calling loudly from the dry reeds or flushing out of the bush.

郭亮 Guo Liang

^ 郭亮 Guo Liang

灰胸竹鸡

灰胸竹鸡行动隐秘，遇人会快速躲进灌丛。在贵州的遵义地区，雨后，车辆刚刚翻过一个山坡，遇到一只在路边觅食的灰胸竹鸡，它快速躲藏，几分钟后探出头来静静地打量着四周。

Bambusicola Thoracicus

The Chinese bamboo partridge (*Bambusicola Thoracicus*) is very secretive; it can swiftly melt into the bush when seeing people. In Zunyi, Guizhou, a car climbed up a mountain slope and saw one looking for food by the roadside after the rain. The partridge immediately hid itself and projected its head out the woods a few minutes later, looking around quietly.

^ 彭建生 Peng Jiansheng

紫水鸡

云南西北部洱海源头的西湖湿地，突然再次发现紫水鸡。紫水鸡被誉为“全世界最美丽的秧鸡”，但在全球范围内遭到捕杀，濒临灭绝。

Porphyrio Porphyrio

Swamphens (*Porphyrio Porphyrio*) were rediscovered in Xihu Wetland at the source region of the Erhai Lake in northwestern Yunnan. It is reputed as the most beautiful swamphen in the world and has been hunted worldwide to near extinction.

董磊 Dong Lei

黑颈鹤

Grus Nigricollis

< 郭亮 Guo Liang

黑颈鹤

冬季的西藏巴松错，这片群山环抱的山谷湖泊区，是很多黑颈鹤的越冬地。湖边的草甸是黑颈鹤的觅食区，每天中午，鹤群会飞来这里觅食，最多的时候，会聚集 50 多只黑颈鹤。黑颈鹤飞越雪山的场景并不多见，尤其是它们迎着雪山随着上升气流飞跃的瞬间。

Grus Nigricollis

Basongcuo, a lake nestling among the hills, is where many black-necked cranes (*Grus Nigricollis*) choose to spend winter. At noon, flocks of up to 50 cranes flew over here, looking for food on the lakeside meadow. Black-necked cranes flying over the snowy mountain was already a rare scene, not to mention the moment when they nosed up against the curve of the mountain under the force of a strong upward air current.

崔林 Cui Lin

大天鹅

河南省三门峡市与黄河对岸的山西省平陆县拥有数万亩的黄河湿地保护区，每年 10 月至次年 3 月，会吸引数万只大天鹅来此栖息越冬。这里也成了观赏及拍摄大天鹅的圣地。慢速追拍的方法既可以拍清楚大天鹅的头颈部，又可以虚化它们扇动的翅膀，再加上适当的摇动，还可以获得“飞行的轨迹”。

Cygnus Cygnus

Sanmenxia City, Henan Province and Pinglu County, Shanxi Province are linked by a Yellow River wetland reserve of several thousand *mu*. From October to March, the reserve is a destination for shutterbugs attracted by thousands of *Cygnus Cygnus* wintering here. Slow-motion shooting can not only capture the head and neck of the whooper swans, but also blur their flapping wings. If the camera swings a little, even the flight paths can be shown.

朱鹮

1981年5月，中国科学院动物研究所鸟类专家刘荫增在陕西省洋县华阳镇，发现了世界上仅有的7只朱鹮。朱鹮是中国自然保护界为数不多的成功案例之一，经过科学家和当地政府20多年的共同努力，如今，朱鹮的数量已经突破了1500只。

Nipponia Nippon

In May 1981, the world's last seven crested ibis (*Nipponia Nippon*) were accidentally discovered by Liu Yinzeng, a bird expert from the China Academy of Sciences Institute of Zoology, in Huayang Town, Yang County, Shaanxi Province. Now, through more than two decades of joint efforts between scientists and the local government, ibis population has increased to over 1,500. This is one of the very few successful cases of China's nature protection projects.

∨ 郭亮 Guo Liang

> 彭建生 Peng Jiansheng

^ 崔林 Cui Lin

胡兀鹫

2011 年初春，在云南西北部的纳帕海湿地，这只胡兀鹫伴随着中国著名藏族歌手亚东的歌曲飞离："默默地向你挥挥手，告别我们轮回的缘分，应召而来天的神鹰，请你带走我一生的荣耀！"

Gypaetus Barbatus

In the early spring of 2011, a bearded vulture (*Gypaetus Barbatus*) flew over the Napahai Wetland in northwestern Yunnan to the singing of Ya Dong, a famous Tibetan singer - "Quietly, I wave to you, bidding adieu to the spirit of samsara; eagles come for my call, please take away the glory of my life."

白尾海雕

白尾海雕每年会迁徙到云南西北部的纳帕海湿地越冬。每天早晨，它们会在高山松上晒够太阳，然后展翅高飞，从空中俯瞰巡视猎物。它们是大型猛禽，需要有足够的猎物才能生存。

Haliaeetus Albicilla

Every year, the white-tailed eagles (*Haliaeetus Albicilla*) migrate to the Napahai Wetland in northwestern Yunnan for the lengthy winter. Every morning, they bask sufficiently on pines before soaring high with outspread wings, overlooking their preys on the ground. These large raptors must have enough preys to survive.

^ 彭建生 Peng Jiansheng

大𫛭

在三江源地区，大𫛭主要以鼠兔为食。因高山草甸树木稀少，大𫛭只能寻找突出的岩石作为落脚点，俯瞰鼠兔的活动。当机会来临时，它们展翅而起，俯冲而下。

Buteo Hemilasius

In Sanjiangyuan, the upland buzzards (*Buteo Hemilasius*) feed themselves mainly on pikas. As the alpine meadow was sparsely wooded, the upland buzzards could only perch on huge rocks to overlook pikas. When the opportunity offered, it spread wings and swooped down.

∨ 郭亮 Guo Liang

灰岩柳莺

2009 年 5 月，IBE 在广西崇左第一次影像调查时被记录到。2010 年研究人员将其确定为新种，种本名 calciatilis，意为“生活在石灰岩中”。

Phylloscopus calciatilis

In May 2009, IBE recorded a kind of warbler in its first image survey in Chongzuo, Guangxi. A year later, the warbler was identified as a new species, limestone leaf warbler (*Phylloscopus calciatilis*).

^ 郭亮 Guo Liang

彭建生 Peng Jiansheng

斑头雁

临近秋季，西藏念青唐古拉山脉下的那曲草原水草丰茂，牛羊成群，一派“风吹草低见牛羊”的景象。青藏高原腹地的候鸟也开始集结，准备飞往更南的雅鲁藏布河谷越冬。

Anser Indicus

The autumn was approaching. On the Nakqu Grassland at the foot of Tanggula Mountain, flocks and herds appeared as grass bended to wind. Migratory birds gathered in the hinterland of the Qinghai-Tibet Plateau, waiting to head further south for the Yarlung Zangbo valley to spend the winter.

< 郭亮 Guo Liang

雪鸽

11月的藏南地区，喜马拉雅山脉与印度接壤的海拔1500米的山谷里，树叶刚刚变黄飘落，晨光唤醒一群雪鸽，它们从海拔近4000米的雪山飞来，在这里歇歇脚。

Columba Leuconota

In November, the leaves just turned yellow and started to fall in southern Tibet. The bright morning lights flooded into a valley that was 1,500 meters above sea level at the border of the Himalayas and India, and aroused a flock of snow pigeons (*Columba Leuconota*). They just took a rest here after a long fly from the snowy mountain at nearly 4,000 meters.

纵纹腹小鸮

秋季的四川石渠地区，天高云淡，阳光非常充足，一只纵纹腹小鸮享受着正午的阳光，同时又观察着鼠兔的活动，长时间地等待猎物，难免有疲倦的时候。

Athene Noctua

The sun shone bright and warm over the autumnal Shiqu. A little owl (*Athene Noctua*) was observing the activities of pikas under the midday sun, feeling tired after the long wait for preys.

< 郭亮 Guo Liang

∨ 郭亮 Guo Liang

^ 郭亮 Guo Liang

大紫胸鹦鹉

Psittacula Derbiana

彭建生 Peng Jiansheng

大紫胸鹦鹉

在梅里雪山山麓的支拉村，上千只大紫胸鹦鹉飞到华山松林觅食。大紫胸鹦鹉是全世界栖息海拔最高的鹦鹉，也是青藏高原特有种。随着青藏高原森林数量日趋减少，如此集大群的场景已经难得一见。

Psittacula Derbiana

In Zhila, a village at the foot of the Mainri Snowy Mountain, thousands of Lord Derby's parakeets (*Psittacula Derbiana*) flew over to the pinus armandii forest, looking for food. Endemic to the Qinghai-Tibet Plateau, they inhabit the highest altitude among all parrot species. The decreasing forests on the Plateau, however, have made large flocks like this a rare phenomenon.

橙翅噪鹛

橙翅噪鹛是中国西南地区常见的一种噪鹛，中国特有。秋天，在雅鲁藏布大峡谷，路边及村庄周围的灌丛经常被它翅上那抹橙黄点亮。

Trochalopteron Elliotii

Elliot's laughingthrush (*Trochalopteron Elliotii*) is a bird species in the Leiothrichidae family. Endemic to China, it is commonly found in the southwest region. In the autumnal Yarlung Zangbo Grand Canyon, the bushes along the roads and surrounding the villages are often dotted with the orange color on its wings.

∨ 郭亮 Guo Liang

^ 郭亮 Guo Liang

灰腹噪鹛

灰腹噪鹛仅生活在中国西藏的东部地区，喜灌丛间跳跃，地面取食。

Garrulax Henrici

Brown-cheeked (*Garrulax Henrici*) laughingthrush is a bush-hopping, ground-foraging bird which lives only in eastern Tibet.

< 郭亮 Guo Liang

画眉

画眉鸟因其华丽的鸣叫而名满中国。在中国广西崇左，一对画眉鸟每天清晨都要经过这片大茅草地，但却羞涩得很，我们难得瞥见其身影。经过 3 天的等待，终于抓拍了一张。

Garrulax Canorus

Chinese hwamei (*Garrulax Canorus*) is hugely popular with its Chinese fans for its attractive and gorgeous song. In Chongzuo, Guangxi, a pair of Hwame is flew over a stretch of weeds every morning, but were too shy to be caught sight of. The final snapshot was worth the three-day wait.

^ 郭亮 Guo Liang

棕草鹛

夏季，中国青海澜沧江源地区，车在山谷中慢行，三四只棕草鹛从车前掠过。它们不时地鸣叫，保持着联络，这只棕草鹛因车辆的阻挡而显得不知所措。棕草鹛是中国特有物种。它们喜欢集结成小群，沿着固定的路线行动。

Ianthocincla Koslowi

It was a nice summer in the source region of the Lancang River. A car was driving slowly along the valley when three or four Tibetan babaxes (*Ianthocincla Koslowi*) skimmed across, which called sometimes to make contacts with each other. A babax seemed rattled about the car. Endemic to China, Tibetan babaxes often follow fixed routes in small groups.

^ 郭亮 Guo Liang

大草鹛

鹛类是中国特有鸟类中的一大类，大草鹛是其中之一。仅在西藏南部可见。

Babax Waddelli

Babblers are endemic birds of China. Among them, Giant babax (*Babax Waddelli*) is seen only in southern Tibet.

^ 郭亮 Guo Liang

朱鹀

藏雀、藏鹀、朱鹀，是青海黄河源地区特有的三种鸟类，也是中国特有的鸟种。朱鹀是中国鸟类雀科朱鹀属下唯一的鸟种，藏雀和藏鹀也都是极为罕见的鸟类。

Urocynchramus Pylzowi

Carpodacus Roborowskii, *Emberiza Koslowi* and *Urocynchramus Pylzowi* are three species of bird endemic to the source region of the Yellow River in Qinghai. *Urocynchramus Pylzowi* belongs to its own family, the Urocynchramidae, of which it is the only member; while *Carpodacus Roborowskii* and *Emberiza Koslowi* are extremely rare.

藏雀

Carpodacus Roborowskii

∨ 郭亮 Guo Liang

藏鹀

Emberiza Koslowi

∨ 郭亮 Guo Liang

^ 董磊 Dong Lei

白腰雪雀

高海拔草原上最常见的鸟类就是白腰雪雀了，它们学会了利用鼠兔的废弃地洞藏身，因此成功地在青藏高原生存下来。这窝小雪雀的父母外出找食物，3个小家伙不肯待在黑暗的地洞里，钻出来站在草地上晒太阳。

Montifringilla Taczanowskii

White rumped snowfinch (*Montifringilla Taczanowskii*) is the most common species of bird on high-altitude grasslands. It has learned to survive the harsh conditions of the Qinghai-Tibet Plateau by hiding in the abandoned underground tunnels of pikas. To get away from the darkness, three baby snowfinches emerged from the tunnel, bathing in the sun rays when their parents were out for food.

∨ 郭亮 Guo Liang

^ 郭亮 Guo Liang

山鹛

山鹛只生活在中国北方地区，零星分布。2 月的医巫闾山，寒风凛冽。一对山鹛躲在山脚下的灌丛中觅食，我们驾驶的车辆打扰了它们。一只羞涩地躲到灌丛深处，另一只好奇地观察着我们。

Rhopophilus Pekinensis

Chinese hill warblers (*Rhopophilus Pekinensis*) are distributed sparsely in northern China only. In the freezing, windswept February of Yiwulü Mountain, a couple of Chinese hill warblersare foraging among the foothill bushes until they are disturbed by our car. One of them hides into the depth of bushes, while the other looks at us with curious eyes.

^ 郭亮 Guo Liang

震旦鸦雀

震旦鸦雀，是中国特有的珍稀鸟种，被称为“鸟中熊猫”。它的名字非常中国化，以古印度人对华夏大地的称呼“震旦”为名。这种鸟的第一个标本采集发现地是中国南京，所以定名为“震旦鸦雀”。

Paradoxornis Heudei

Paradoxornis Heudei, a species of bird endemic to China, is even referred to as the "panda in the birds" for its rarity. It has a special Chinese name which literally means "Zhendan (the ancient Indian name for China) crow finch" to honor the Chinese city Nanjing where its first specimen was discovered.

^ 张瑜 Zhang Yu

楼燕

楼燕是世界上平飞速度数一数二的鸟，一生中除了孵卵育雏几乎都在空中飘荡。在北京，许多古建筑房檐下的孔隙里都曾有楼燕繁殖，不过近些年来，伴随着这些建筑维修安装防鸟网，它们的日子变得越来越不好过了。

Apus

Common swift (*Apus*), the most aerial of birds renowned for its superb level speed, spends its life almost entirely on the wings except when incubating and brooding. In Beijing, many old buildings had holes under eaves where swifts used to nest. But days keep getting tougher for the swifts as more and more building are now equipped with bird-proof nets.

植物
Plants

藿香叶绿绒蒿

在西藏东部喜马拉雅山区南迦巴瓦山脚下，发现了一丛开得正盛的藿香叶绿绒蒿。这种高原植物只在夏季的藏东南区域开放，强烈的阳光下，半透明的蓝色花瓣在逆光下特别漂亮，带有传统中国水墨画的笔法。

Meconopsis Betonicifolia

At the foot of Namcha Barwa, a Himalaya peak in eastern Tibet, I fell in with a blossoming Himalayan blue poppy (*Meconopsis Betonicifolia*). The plateau plant blooms only in southeastern Tibet in summer. Against a glaring backlight, its translucent blue petals exude a certain of Chinese ink-wash paintings.

< 彭建生 Peng Jiansheng

红花绿绒蒿

红花绿绒蒿生长在海拔 2800~4500 米的高山上，叶基生单葶花絮，花葶以柔软的糙毛御寒。这是一株双葶花絮的红花绿绒蒿，它们相依相偎，含苞待放，在花葶下部有昆虫即将造访。

Meconopsis Punicea

Meconopsis Punicea grows in mountains with an elevation of 2,800-4,500m. It usually has a single-headed hispid inflorescence arising from the base of a leaf. This plant, however, has a double-headed inflorescence, below which an insect is anxiously waiting for its budding.

< 徐健 Xu Jian

红花绿绒蒿

红花绿绒蒿是中国特有种，生长在四川西北部、西藏东北部、青海东南部和甘肃西南部的高山草甸上，是一种非常美丽的高山植物。在四川石渠地区的一个山谷里，这棵红花绿绒蒿正在阳光下盛放。

Meconopsis Punicea

In a valley in Sêrxü County, Sichuan, I came across a *Meconopsis Punicea* blossoming in the sun. The beautiful species endemic to China grows on alpine meadows in northwestern Sichuan, northeastern Tibet, southeastern Qinghai and southwestern Gansu.

单叶绿绒蒿

单叶绿绒蒿恰好是整个绿绒蒿家族进化、演变的一个特殊例证，它没有真正的茎，挺立在地面上的只是它的花梗，每一枝花梗上只有一朵花。它们生活在林下、灌丛、高山草地或石缝中，和原始绿绒蒿的生活环境有几分相似——虽然爬上了高海拔地区，却并没有真正脱离其他草木所营造的小环境。

Meconopsis Simplicifolia

Meconopsis Simplicifolia is a special case of the evolutionary history of meconopsis. Without a stem but having a pedicel, the plant lives in the undergrowth, coppices, alpine meadows, and clefts of the rock. Although it has climbed to the uplands, it still relies on the micro-environment created by other plants as its ancestors did.

∨ 彭建生 Peng Jiansheng

∧ 彭建生 Peng Jiansheng

刺瓣绿绒蒿

绝大部分绿绒蒿都是多年生一次开花就结束生命，它们积蓄多年的养分，在七八月高原的蓝天下努力绽放以吸引昆虫造访。为了保证生命传递，刺瓣绿绒蒿全身被满坚硬的毛刺，甚至在它的花瓣上也有毛刺，以防其他动物啃食。高原上的生命何其短暂，又何其艰难、顽强！

Meconopsis Racemosa

Almost all members of the meconopsis genus are alpine perennials which die after their first flowering. With nutrients stored for years, they bloom on a sunny summer day and wait for the visits of insects. To protect itself from other animals, *Meconopsis racemosa* has stiff burrs all over its body (even on its petals). So tenacious and transient are lives on the plateau.

> 牛洋 Niu Yang

全缘叶绿绒蒿

全缘叶绿绒蒿是绿绒蒿属植物中分布较广、容易见到的一员，但滇西北的全缘叶绿绒蒿却有着独特的魅力。在这里，它们常植根于因大火而烧枯的柏树灌丛中，在虬曲的枯枝中绽放。自然的法则在这样的画面中得以奇妙地体现。

Meconopsis Integrifolia

Meconopsis Integrifolia is a widespread species of the Meconopsis genus, yet its ramifications in northwestern Yunnan has their unique charm. Here, they often take roots in cypresses burnt to the ground, and miraculously stick their neck out of twisted branches.

^ 彭建生 Peng Jiansheng

云南杜鹃

云南滇西北的澜沧江河谷，温润的山风催促着百花开放，云南杜鹃奋勇当先，它高抬枝头绽放，有时候可以把整片山林都染上色彩，犹如献给神山卡瓦格博的鲜花供奉。

Rhododendron Yunnanense

In Lancang River Valley in northwestern Yunnan, moist mountain breeze is hastening the blooming of various plants. Towering above them is the widespread *Rhododendron Yunnanense* offering up its bright blossoms to the sacred Kawa Garbo Peak.

< 牛洋 Niu Yang

栎叶杜鹃

丽江老君山简直是由杜鹃属植物组成的天然花园，一株巨大的栎叶杜鹃树令人臣服，广角镜头记录下了密密匝匝、层层叠叠的花簇。

Rhododendron Phaeochrysum

Laojun Mountain in Lijiang is literally a natural garden of rhododendrons. My wide-angle lens captured at owering *Rhododendron Phaeochrysum* with dense layers of inflorescences.

∨ 董磊 Dong Lei

报春花

报春花是中国高原四大名花之一，与绿绒蒿、杜鹃、龙胆齐名。2011 年夏天，我再次看见了这种开黄花的锡金报春，我选择了一棵枯树干前面最整齐的一丛，仰视拍摄。自然界里哪怕是再渺小的生物，也都蕴含着巨大的生命力。

Primula Sp.

Primroses (*Primula Sp.*), together with Himalayan blue poppies, rhododendrons and gentians, represent the top four alpine bloomers in China. In summer 2011, I had another encounter with *Primula Sp.* My low-angle shots captured a neat cluster of the white flowers against a withered tree trunk. Even the tiniest creatures can express an exuberant vitality.

< 左凌仁 Zuo Lingren

报春花

一场突如其来的大雨，让青海玛多星星海湿地内刚刚绽放的报春花显得更加艳丽。一丛丛的报春花绵延不绝，沾满雨水的草甸和报春花晶莹剔透，漂亮极了，这种如梦似幻的场景实属难得一见。

Primula Sp.

An unforeseen rain adds beauty to the fresh blossoms of *Primula Sp.* in Xingxinghai Wetland, Madoi County, Qinghai. It is rare to see so many fairy queens dancing on a rain-soaked meadow.

^ 彭建生 Peng Jiansheng

美丽乌头

我们在香格里拉大雪山一个海拔 4500 米的山头做生态监测时，一场突如其来的大雪把人赶下山，途中遇到这株美丽乌头，它身披冰雪，宛若娇羞的仙子。

Aconitum Pulchellum

We chanced upon a delicate, snow-covered *Aconitum pulchellum* after a heavy snowfall suspended our eco-monitoring project and drove us down a 4,500m peak.

^ 彭建生 Peng Jiansheng

珠芽蓼

2013年9月，我们前往大雪山从事高山生态监测。12日，我们攀爬到海拔4200米的山顶，正在为高山灌丛检测样品，突然天降暴雪，迫使我们迅速下撤。途中一株鲜红的珠芽蓼依然傲雪挺立。珠芽蓼是蓼科蓼属的优质牧草，是牦牛的主要食物。

Polygonum Viviparum

September 2013 found me engaged in an eco-monitoring project in Daxue Mountains. On Sept.12th, a snow-capped, scarlet *Polygonum Viviparum* caught our sight after an unexpected snowstorm halted our assessment of alpine coppice quadrats and forced us down a 4,200m peak. A member of bistorta genus, polygonaceae family, the plant is a quality source of forage and the staple diet of yaks.

^ 牛洋 Niu Yang

红花岩梅

红花岩梅是岩梅科的一员，生长在高海拔的石崖或草甸上。它们是一类低矮的垫状植物，个体呈半球形，叶片小而密集，这样的形态有利于其在高山之巅减少寒冷和强风的伤害。

Diapensia Purpurea

A member of Diapensiaceae family, *Diapensia Purpurea* grows on high-elevation rock cliffs or meadows. The hemispheric cushion plant has dense small leaves resistant to cold and strong winds.

< 董磊 Dong Lei

龙胆

在一天里阳光最强烈的时候，青海三江源地区的草原上的龙胆花才会将花瓣打开。龙胆是青藏高原上一种很有特点的植物，因为花期接近秋天，而且青藏高原温差特别大，龙胆因此在花期进化出了无太阳时闭合、有太阳时绽放的习性。

Gentiana Sp.

Gentiana Sp. is a unique plant on the steppes of Qinghai-Tibet Plateau. Owing to huge temperature differences in late summer, its petals close at night and unfold only in the brightest sunshine.

> 牛洋 Niu Yang

新疆猪牙花

在新疆阿勒泰地区稀疏的落叶松林中，5 月的早晨，寒意未尽，猪牙花在清冽的空气中绽放，反卷的花瓣好似天使的羽翼。

Erythronium Sibiricum

On a chilly, fresh May morning, *Erythronium Sibiricum* is blooming in a sparse larch forest in Altay, Xinjiang. Its inverted tepals look like the wings of an angel.

^ 牛洋 Niu Yang

澜沧雪灵芝

5月，低地已进入初夏。滇西北高山流石滩上，季节性的积雪尚未融化。澜沧雪灵芝绽放出白色的小花，云雾雀儿豆也开出紫红色的花朵——高山流石滩的春天才刚刚开始，雪山花园正要进入最美丽繁盛的季节。

Arenaria Lancangensis

In May, when the lowlands are bracing for the first heat wave of summer, alpine screes in northwestern Yunnan have just darkened the doorstep of spring. Here, seasonal snow is dotted with little white blossoms of *Arenaria Lancangensis* and magenta flowers of Chesneya nubigena. The prime of the year is coming to the snow mountain garden.

^ 朱仁斌 Zhu Renbin

美丽肋柱花

肋柱花属植物的雌蕊没有花柱，柱头沿着子房的缝合线下延，形如肋，由此得名。美丽肋柱花为一年生植物，只分布于我国湖北、陕西、四川和云南。每年八九月，陕西太白山海拔 3200 米左右的地方就会盛开这种花朵。

Lomatogonium Bellum

The pistil of *Lomatogonium* is without style. The stigma extends downward along the ventral suture of the ovary and takes the shape of a rib. *Lomatogonium Bellum* is an annual plant which grows exclusively in China (in Hubei, Shaanxi, Sichuan and Yunnan). In August and September each year, it blooms at an elevation of 3,200m on Taibai Mountain in Shaanxi.

^ 徐健 Xu Jian

大叶木莲

大叶木莲喜欢生长在阴湿的沟谷两旁或山沟低洼处，它们的叶片大而浓密，花朵硕大而美丽。目前，大叶木莲成熟个体野外仅存 200 余株，面临野外灭绝的威胁。在其主要分布地滇东南的火烧梁子和古林箐，已经建立了保护区。

Manglietia Megaphylla

Manglietia Megaphylla grows well in damp and dark valleys or other low-lying areas. It has big, thick leaves and large, beautiful flowers. Currently, there are only 200-odd mature individuals of *Manglietia Megaphylla* in the wild. The plant is literally on the verge of extinction. Natural reserves have been established at its main habitats - Huoshaoliangzi and Gulinqing in southeastern Yunnan.

塔黄

2012 年 6 月，在滇西北一处海拔 4500 米左右的高山流石滩附近，从很远就可以看到生长在这里的"巨人"——塔黄，身披硕大的浅黄色苞片。一种很小的蚊类被它的颜色和所释放出的特殊气味吸引，为它完成传粉。这种巨大的植物一生只有一次机会开花结果，但只有很少的种子有机会再次成长为一个"巨人"。

Rheum Nobile

In June 2012, I descried a *Rheum Nobile* at 4,500m scree in northwestern Yunnan. It looked huge even from distance. A tiny mosquito was pollinating the plant, apparently attracted by its yellow bracts and special scent. *Rheum Nobile* only has one chance to bloom and bear fruit throughout its lifetime, yet few of its seeds will grow into another giant.

∨ 牛洋 Niu Yang

吴双 Wu Shuang
中国广西壮族自治区靖西县
Jingxi County, Guangxi

广西边陲靖西县有着典型的喀斯特地貌，很多河流里都生长着海菜花的变种——靖西海菜花，它们犹如白色的精灵点缀着这里的秀美山水。该县“绣球之乡”旧州古镇附近的山形陡峻，倒映在漂浮着靖西海菜花的鹅泉河上。

Jingxi, a border county in Guangxi, features typical karst topography. In rivers like Equan, whose serene water reflects the precipitous ridges near Jiuzhou, an "ancient town of big-leaf hydrangea", grows an elf-like three-petalwhite flower named after the county.

^ 崔林 Cui Lin

中国吉林省东部的珲春，属于中温带海洋性季风气候。这里群山环抱，遍布针叶林及针叶阔叶混交林，森林覆盖率很高，良好的自然条件为野生动物提供了繁衍生息的栖息地。此时正是 10 月中秋，层林尽染，秋意正浓。

Hunchunin eastern Jilin features a cool-temperate oceanic monsoon climate. The city is densely covered with (mixed) coniferous forests. Such natural conditions are ideal for wildlife to live and multiply. Now in the middle of October, a fantasia of tableaux is playing among the thickly-wooded hills surrounding the city.

^ 徐健 Xu Jian

望天树

望天树是云南西双版纳热带雨林的象征，树高可达七八十米，甚至更高，是名副其实的热带雨林“巨人”，站在“巨人”脚下，仰望直指蓝天的巨树，身心都受到洗礼，让人不由自主地赞叹自然的伟大。

Parashorea Cathayensis

Parashorea Cathayensis is a symbol of the tropical rainforests in Xishuangbanna, Yunnan. With a height of 70-80 meters or more, the tree is arguably a "crane among chickens". Standing under the giant umbrella, I cannot but be amazed at the grandeur of nature.

程斌 Cheng Bin

西藏东南部的墨脱地区，位于中印边境，长期以来无路可通，与世隔绝，那里的自然生态环境是目前中国最好的，无处能及。

Mêdog is an out-of-the-way county on the China-Indian border in southeastern Tibet. Its natural environment is by far the best in China.

昆虫和蜘蛛
Insects and Spiders

徐廷程 Xu Tingcheng

螽斯科

蜕皮时是昆虫最脆弱的时候，不但对天敌没有任何防御办法，甚至连逃跑的能力都没有。因此很多昆虫都会选择夜间蜕皮避开捕食者。夜拍时见到一只正在蜕皮的螽斯，它的蜕皮过程已经接近尾声，接下来只需静静地等待翅膀晾干。

Tettigoniidae

Insects are most vulnerable during sloughing - incapable of fleeing, they are easy prey for their natural enemies. Many insects, therefore, shed their skin during the night to avoid predators. My night shots captured a creature which was almost through the sloughing process and patiently waiting for its wings to dry.

< 雷波 Lei Bo

乌微桥原蟌

盛夏的小溪边，一直有乌微桥原蟌在飞舞，当我看到阳光洒在溪水上的时候，我很想用镜头展现出乌微桥原蟌在洒满阳光的水面上飞舞的场景，经过一段时间耐心的等待，终于等来了我想要的画面。

Prodasineura Autumnalis

Patience repaid me with the coveted view of *Prodasineura Autumnalis* dancing over a sparkling brook in summer.

毛狭扇蟌

夜间，毛狭扇蟌躲藏在水边的草丛深处休息，透过树叶的叶洞，它的眼睛里依旧透露着一丝警惕和不安。

Copera Ciliate

A *Copera Ciliate* is resting deep in the waterside grass at night. Its uneasy, wary eyes are peeping through a leaf hole.

雷波 Lei Bo

< 程斌 Cheng Bin

君主绢蝶

明媚的阳光下，一只君主绢蝶像滑翔机一样乘着上升的热气流巡视着大地。它们的翅膀看上去如同半透明的丝绢，故名“绢蝶”。绢蝶是世界上最早被立法保护的昆虫，全世界总共有 50~53 种，中国的绢蝶种类占已知种类的三分之二还多。

Parnassius Imperator

A *Parnassius Imperator* is floating with the hot upward current in bright sunshine, overlooking the terra firma. Its wings look like translucent silk handkerchiefs. The butterfly is the first insect to be protected by law. China accounts for over two-thirds of the 50-53 known subspecies around the world.

∨ 袁明辉 Yuan Minghui

曲纹紫灰蝶

藤蔓上的这对曲纹紫灰蝶正在交尾。曲纹紫灰蝶也被称为双头蝴蝶，其实这是后翅尾突表现出来的障眼法。曲纹紫灰蝶翅膀底部有着相似的眼状斑纹，这掩人耳目的触角和头部很容易被误认为是头部，使之能够在紧急时刻避开危险。

Chilades Pandava

A pair of *Chilades Pandava* are mating on a vine. To keep predators at bay, the butterfly has a false head consisting of hind wing tail and similar eyespots.

^ 程斌 Cheng Bin

丽眼斑螳

昆虫在亿万年的演化史中，演化出了一些独特的生态行为。这种演化经常会让人们赞叹不已。图中的这种螳螂若虫，像是掉落在叶子上的花蕊，静止的时候，它的整个身体会和花朵融为一体。

Creobroter Gemmata

Several billion years of evolution has endowed insects with some amazingly unique behavior. Like a bud fallen on a leaf, this mantis nymph, in its stillness, co-exists harmoniously with the flower.

兰花螳螂

兰花螳螂，以身形似兰花得名，这只末龄的兰花螳螂在阳光下，更加显得如鲜花般娇艳。

Hymenopus Sp.

Hymenopus Sp. in the larva state resembles an orchid. This one is, however, in its twilight years. It looks even more like a delicate flower in the sunlight.

∨ 范毅 Fan Yi

< 雷波 Lei Bo

斑翅始螳

在云南孟连的亚热带森林中，在一条不起眼的小路边，发现了这只斑翅始螳。它潜伏在绿叶底下伺机捕食，华贵的形态和美妙的斑纹打破了叶脉线条的节奏，让周围的一切显得融和而宁静。

Eomantis Guttatipennis

In the subtropical forest of Menglian, Yunnan, I noticed an *Eomantis Guttatipennis* on an inconspicuous path. The insect was lurking under the leaves, ready to hunt. The beautiful stripes on its elegant body stood out from the leaves. Everything was so harmonious and peaceful.

^ 程斌 Cheng Bin

棉红蝽

蝽在春夏之交的繁殖季节经常会聚集在一起，它们几乎无处不在，不仅在山林中异常活跃，甚至在人类的居住环境中也是常客。棉红蝽钻进掉落在地面的木棉果实的棉絮中，用刺吸式口器，像吸管一样扎进木棉果实中，吸食营养物质。

Dvsdercus Cinqulatus

Stinkbugs will group together during the breeding season at the turn of spring and summer. They are practically everywhere - not only in forests, but also in populated places. *Dvsdercus Cinqulatus* will worm its way into the cotton wool inside the fallen fruit of cotton trees, and withdraw nutrients by means of its piercing-sucking mouthpart.

^ 程斌 Cheng Bin

中华蜜蜂

中华蜜蜂体形较小，是中国独有的蜜蜂。以杂木树为主的森林群落及传统农业作物主要靠它们传粉。

Apis Cerana

The eastern honeybee (*Apis Cerana*) is a small-sized bee endemic to China. It is the main pollinator in miscellaneous forests and traditional agricultural landscapes.

< 雷波 Lei Bo

东亚无垫蜂

经过一段时间的观察，我找到了东亚无垫蜂的临时聚集处。每到夜幕降临之际，它们便从四面八方赶来，一只一只地咬在纤细的枯枝上，一起度过夜晚。

Amegilla Parhypate

After a period of observation, I discovered a temporary gathering place of *Amegilla Parhypate*. At the close of every evening, the insects would come here from all sides and cluster around withered branches.

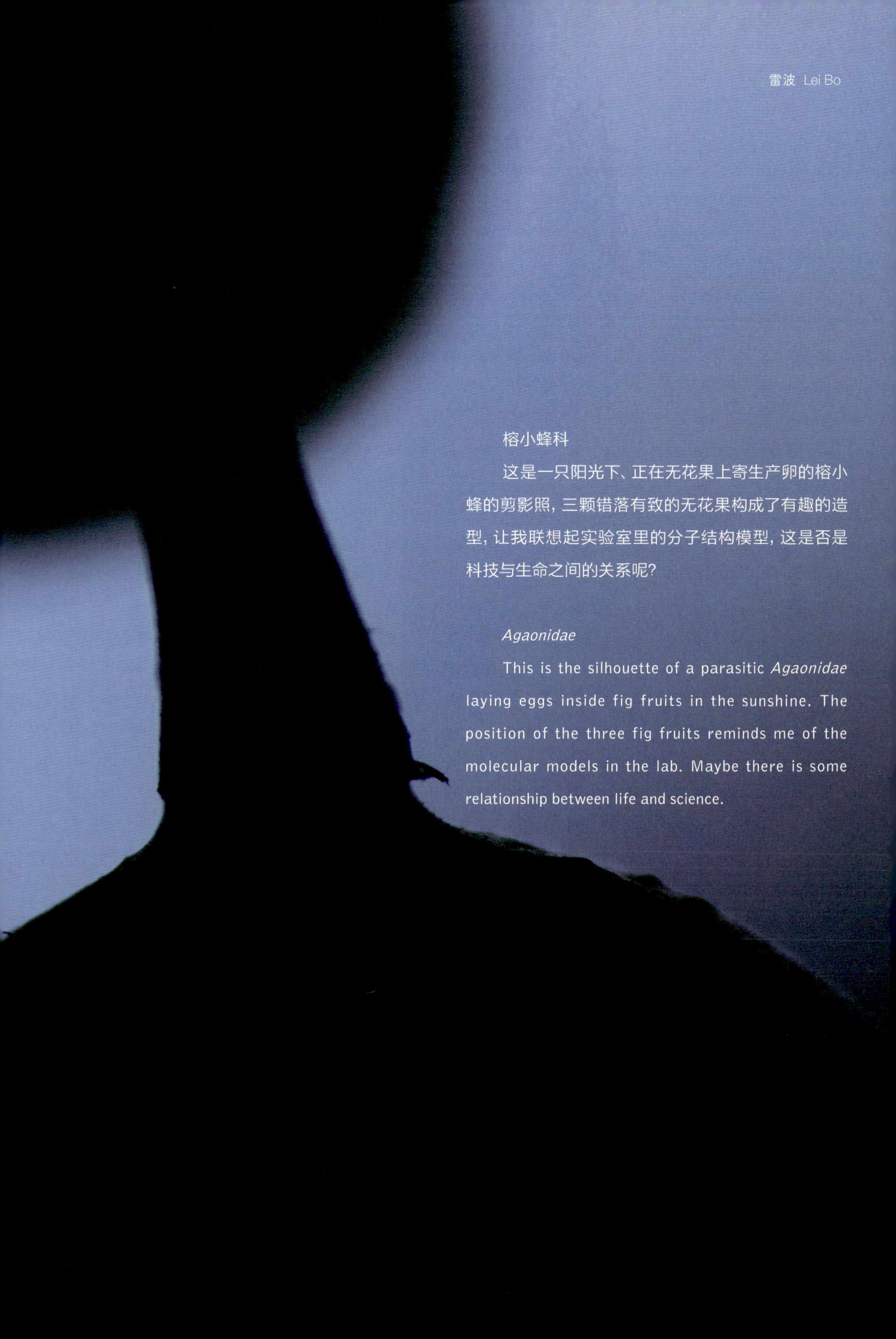

雷波 Lei Bo

榕小蜂科

这是一只阳光下、正在无花果上寄生产卵的榕小蜂的剪影照，三颗错落有致的无花果构成了有趣的造型，让我联想起实验室里的分子结构模型，这是否是科技与生命之间的关系呢?

Agaonidae

This is the silhouette of a parasitic *Agaonidae* laying eggs inside fig fruits in the sunshine. The position of the three fig fruits reminds me of the molecular models in the lab. Maybe there is some relationship between life and science.

雷波 Lei Bo

大蚊

躲藏在暗处的大蚊，它们通过不断晃动身体来达到隐身的目的，就像风扇的扇叶，快速转动的时候你看不清楚，甚至什么也看不见，我用表达运动轨迹的手法，来表现这种奇特的躲避敌害的方式，乍一看，还真像一件舞动的艺术品！

Tipulidae

Hiding in the dark, a *Tipulidae* will constantly shake its body to make itself invisible - just like the fan blades, which are hardly visible when rotating. Our trajectory reproduces the peculiar way of the elegant dancer to avoid enemies.

滇叶䗛

在云南普洱，我终于发现了传言中的滇叶䗛，这种奇特的生物模拟出和树叶一模一样的形态，但我却苦于如何将其表达出来。在经过一段时间的思考之后，我用星轨来表现时间的流逝，以赞美千百年进化的奇迹！

Phyllium Yunnanense

In Pu'er, Yunnan, I eventually found the legendary insect of *Phyllium Yunnanense*. Its ability to mimic tree leaves beggars description. My star trail shots are an ode to the miracles of evolution.

∨ 雷波 Lei Bo

> 刘彦鸣 Liu Yanming

象白蚁

这是在建的白蚁巢口,“长鼻子”就是它们的兵器。这个“长鼻子”其实是化学武器喷管,常用喷胶的方法进行防卫。象白蚁身体柔软,但抵挡不住它们的天敌蚂蚁的机械式撕咬,于是演化出不用近身搏斗就能给节肢动物体表造成大面积烧伤的防卫术。

Nasutitermes Sp.

This is a nest of *Nasutitermes Sp.* under construction. Too delicate to withstand the stings of ants, its natural enemies, the termite uses its long nose as a chemical weapon. The glue sprayed by the nose can cause extensive burns to arthropods, thus obviating the need for close-quarters combat.

条纹绿蟹蛛

蟹蛛的种类很多，它们的共同点是能够让自己拟态隐形，与环境完美融合，有的拟态为花蕊，有的几乎透明，就像这只条纹绿蟹蛛。“隐形”之后，它们既能够让猎物对自己视而不见，也可以捕猎比自己体形大许多倍的猎物。

Oxytate Striatipes

Mimicry skills are the common denominator of the big family of crab spiders. Some of them can pretend to be buds; some, like the *Oxytate Striatipes*, can turn transparent and catch their quarry off guard. Crab spiders are able to hunt animals several times their size.

∨ 程斌 Cheng Bin

蛛形纲

虽然多数蜘蛛都是依靠织不同的网进行捕猎的，但也有例外，有些种类是游走型的猎手，它们不仅依靠动作快捷，也依赖于拟态。这种和白色半透明的菌类协同演化的蜘蛛，体长只有 3 毫米，如果它们钻到这片真菌中，瞬间就能隐身不见。

Arachnoidea

Although most spiders hunt with nets, some are agile guerrillas good at mimicry. The spiders I found blended well with the translucent white surroundings. Measuring only 3mm in length, they were nearly invisible among the fungi.

∨ 程斌 Cheng Bin

两爬和鱼类
Amphibians，reptiles and fish

< 程斌 Cheng Bin

树蛙科蝌蚪

树蛙与人们印象中常见的蛙类不同，它们不仅拥有很好的跳跃能力，四肢上的吸盘还能让它们在植物上任意活动。与青蛙不同，树蛙的卵多数产在水面以上的植物枝茎上，密密的泡沫保持着水分，择机孵化。如今，很多种树蛙的栖息地在消失。

Rhacophoridae

The *Rhacophoridae* are different to common frogs. Not only are they terrific jumpers, but they can also move about freely on plants by means of sucking discs on the arms and legs. Unlike their peers, the *Rhacophoridae* lay their foam-covered eggs in branches and stems above the water surface. Such eggs will hatch in due course. To our regret, many subspecies of *Rhacophoridae* are losing their habitats.

^ 范毅 Fan Yi

原刺髭蟾

美丽的大围山保护区，一年中有 300 天都在大雾之中，而这些大雾中生活着原刺髭蟾这种原始而古老的两栖类动物。它们行动缓慢，没有一般蛙类的跳跃能力，只能靠隐藏的方式，来争取生命的延续。

Vibrissaphora Promustache

In Dawei Mountain, a natural reserve perennially shrouded in heavy mists, lives the ancient primitive amphibian. Unable to jump like other frogs, *Vibrissaphora Promustache* is dragging out a sluggish existence by hiding.

^ 侯勉 Hou Mian

陇川小树蛙

夜间，陇川小树蛙的雄蛙会站在阔叶灌木的顶端演唱优美的夜曲，以吸引雌性的注意。陇川小树蛙是一个神秘的物种，它们并不在水边生活，一生都在潮湿的灌丛中度过。自确认有这个物种以来，一直没有有关它的清晰的野外照片，更不用说鸣叫状态下的生态照了，能够见到并记录下它的这种行为，实在很幸运。

Philautus Longchuanensis

Philautus Longchuanensis is a mysterious frog species. It does not live by the water, but in moist coppices. At night, male *Philautus Longchuanensis* will stand on top of broadleaf shrubs to "serenade" potential mates. Up till now, there was not a single clear picture of the frog in the wild - still less a croaking one - since its name was officially published, so it is indeed fortunate to witness and record its behavior.

变色树蜥

在中国华南地区，变色树蜥很常见，甚至在街边的绿化带植物上也可以发现它们的身影。作为冷血动物，它们经常需要爬到高处享受温暖的阳光。这只正处于繁殖期的雄性变色树蜥，前半身呈现出鲜艳醒目的朱红色，这表明它正在求偶。

Calotes Versicolor

Calotes Versicolor is ubiquitous in South China - it may feel at home even among roadside plants. As an ectotherm, it often needs climbing to higher places to bath in the sunshine. This male *Calotes Versicolor* has put on a vermillion shirt, trying to draw attention from potential lovers.

∨ 郭亮 Guo Liang

雷波 Lei Bo

丽棘蜥

雨后的树林里，凉风中透着一丝安静。这只藏于树后的丽棘蜥似乎和我一样在享受着盛夏中难得的一刻清凉。即使我把相机凑得很近，它也毫不在意。它真是树林中的精灵！

Acanthosaura Lepidogaster

An *Acanthosaura Lepidogaster* is resting behind a tree, enjoying the rare coolness and peace of the woods after a midsummer rain. It does not even care about my moving the camera closer. It indeed is an elf of the woods.

凭祥睑虎 *Goniurosaurus Luii*

嘉道理睑虎 *Goniurosaurus Kadoorieorum*

睑虎

睑虎因有可活动的眼睑而得名。全世界已发现的睑虎共 17 种。中国拥有其中 10 种，是世界上睑虎种类最丰富的国家。它们大多栖息于喀斯特或火山岩地貌环境中，因其外表华丽而成了宠物市场上炙手可热的商品。商业采集和栖息地破坏使得睑虎的野外种群数量急剧下降。从 2014 年开始，我致力为它们留下尽可能多的影像资料。在四年的拍摄中，我深切地体会到这些并非明星物种的小家伙们的保护形势岌岌可危。在一些往年拍摄的地点，已经越来越难见到它们的身影。希望这些影像资料能让更多的人对这些默默无闻的洞穴精灵给予关注。

霸王岭睑虎 *Goniurosaurus Bawanglingensis*

广西睑虎 *Goniurosaurus Kwangsiensis*

Goniurosaurus

Goniurosaurus is a genus of geckos known for their movable eyelids. Of the 17 Goniurosaurus species around the world, a staggering ten can be found in China. Most of the good-looking geckos live in karst or lava landscapes. Unfortunately, their wild population is plummeting due to hunting for the pet trade and habitat destruction. Since 2014, I have been trying to take as many photos as possible for the geckos. Four years on, I am well aware how overlooked and endangered they are. Even in their previous year haunts, the geckos are harder to find. These photos, I hope, will draw more attention to the underappreciated "elves of the caves".

红斑高山蝮

2012年，IBE影像调查在长江源头通天河流域拍摄到的高原蝮蛇。研究人员依此影像记录图片和拍摄地点，于2016年再次找到。研究人员发现这是一种高原蝮蛇新种，并将其命名为“红斑高山蝮”。

Gloydius Rubromaculatus

During an image survey in 2012, IBE captured on camera a kind of alpine pit viper in Tongtian River basin at the source of the Yangtze River. Four years later, the very kind of viper was spotted by researchers in the same area. It was later identified as a new species, red-spotted (*Gloydius Rubromaculatus*) alpine pit viper.

翠青蛇

透过叶片的空隙，看到有一条正在伏击狩猎的翠青蛇。许多小型生物，都在演化过程中选择了绿色，以便与环境融为一体。古老的蛇类是进化非常成功的一类生物，线条状的身体和鳞片，让它们几乎可以在任何环境里生存。

Eurypholis Major

Many small creatures, like the ambushing greater green snake (*Eurypholis Major*) I saw through the leaves, developed green skins over time to blend into the surroundings. Long-term evolutionary success endows snakes with an elongated body and scales fit for almost any environment.

∨ 程斌 Cheng Bin

扬子鳄

扬子鳄是一种小型鳄鱼，是中国特有的珍稀物种。中国政府通过近20余年的努力，使得扬子鳄的人工繁殖数量超过了10000条，不过野外的种群数量仅剩不到130条，这条扬子鳄是在浙江长兴自然保护区半人工环境下养殖的。

Alligator Sinensis

The Yangtze alligator (*Alligator Sinensis*) is a rare small alligator endemic to China. Through 20-odd years of effort, its total captive population has topped 10,000. However, there are fewer than 130 Yangtze alligators in the wild. This alligator is bred at Changxing Nature Reserve, a semi-artificial reproduction center.

∨ 徐健 Xu Jian

∨ 侯勉 Hou Mian

湄公鱊

这是一种神秘的小鱼，国内还没有正式记录，过去国内将它错认为短须鱊，但是它与短须鱊的分布区域、栖息环境和外观特征完全不同。其本种原记录生存于越南和老挝北部的湄公河中游水系，在中国仅见于云南南部普洱市与西双版纳州的部分热带溪流或小河。我偶然从驴友在西双版纳热带雨林中徒步时拍的图片里发现了它的倩影，然后一路追寻找到热带雨林深处的那条溪流，并以当地原生态为背景拍摄了这种美丽飘逸的小鱼。后来经过仔细查核，发现它不是原记录的短须鱊，其综合特征与湄公鱊的描述及分布完全一致，才确定它是国产淡水鱼类的新记录——湄公鱊。

Acheilognathus Deignani

Acheilognathus Deignani is a mysterious small fish of which China had no official record. It used to be mistaken for *Acheilognathus Barbatulus*, a species which lives in other areas and looks completely different. *Acheilognathus Deignani* was originally found in the middle stream of Mekong River in Vietnam and northern Laos. In China, it has only been seen in tropical brooks and rivers in Pu'er and Xishuangbanna in southern Yunnan. On one occasion, I caught a glimpse of the obscure species in photos taken by backpackers in the tropical rainforests of Xishuangbanna. Following their footsteps, I pinpointed the very brook where it was found and then photograph its beautiful andelegant shape against the pristine backdrop. Careful examination of its general characteristics convinced me that what I found not only was not *Acheilognathus Barbatulus*, but also filled a gap in China's record of freshwater fish.

^ 程斌 Cheng Bin

青海湖裸鲤

因为地处偏远的藏西北边境地带，阿里地区人口非常稀少，这里有藏传佛教文化中的神山和圣湖，许多生物世代生活在这里。这里有众多的湖泊和发达的水系，生活着一类名为裸鲤的鱼类，受益于当地“不吃鱼”的文化而得以生存。

Gymnocypris Przewalskii

Since eating fish is prohibited by Tibetan culture, creatures like *Gymnocypris Przewalskii* manage to live in the sacred mountains and lakes of Ngari, a remote and sparsely populated prefecture on the northwest border of Tibet.

水下生物

Underwater creatures

∧ 吴立新 Wu Lixin

斑点鹰鳐

斑点鹰鳐是鲨鱼的近亲，但在它身上丝毫看不到掠食者的狰狞。它优雅的游动姿态，像空中翱翔的雄鹰。这些物种其实一直远离人们的视线，不为大众所熟知。

Spotted Eagle Ray

A close relative as it is of the shark, *Spotted Eagle Ray* is not in any way a ferocious predator. In fact, it swims as elegantly as an eagle. The lesser known species leads a low-key life at Xisha Islands.

< 吴立新 Wu Lixin

海底柏柳珊瑚

中国南海的西沙群岛，柳珊瑚会“落地生根”，附着在珊瑚礁海底峭壁上，在这种洋流相对强劲的地方，它们会用触手捕捉被海流带来的浮游生物。

Melithaea Sp.

Melithaea Sp., which covers the seamounts of Xisha Islands, uses its tentacles to catch plankton brought by strong currents.

桌状轴孔珊瑚

南沙群岛的桌状珊瑚，是热带珊瑚礁常见的生长相对较快的一种造礁石珊瑚。不同种类以及不同个体的珊瑚礁之间会相互竞争，甚至相互蚕食，桌状珊瑚是其中典型的代表物种。

Acropora Hyacinthus

Acropora Hyacinthus at Nansha Islands is a common, fast-growing hermatypic coral. Different individuals and different kinds of corals will compete with, or even eat into, each other.

∨ 吴立新 Wu Lixin

∧ 吴立新 Wu Lixin

越前水母

越前水母，俗称绵蜇、沙蜇，每年夏末，会在东海、黄海大量出现，成年个体最大会长到直径 1 米左右。在能见度较差的温带海域黄海北部的獐子岛海域拍摄时，需要尽可能地靠近它们，还要不时地防备其触手。

Neopilema Nomurai

At the end of each summer, the East China Sea and the Yellow Sea will teem with *Neopilema Nomurai*. The maximum diameter of a fully grown *Neopilema Nomurai* is about one meter. Taking pictures of the jellyfish at the turbid waters around Zhangzi Island, we have to stay as close to it as possible while keeping an eye on the tentacles.

影像生物调查所（IBE）简介
Introduction to IBE

影像生物调查所（IBE）是一家由职业摄影师创立的自然影像机构。自 2009 年创立起，IBE 就用一种颇创新精神的生物调查方法开展工作：IBE 的每次调查都会收集带有科学数据的影像作为证据来进行生物多样性究；每次调查均由多位专业自然摄影师合作完成，他们分工拍摄不同生物类群，之后将影像、科学数据汇总成调查成果。

IBE 目前的主要业务内容包括：影像生物多样性调查、自然纪录片拍摄制作、自然影像展览、自然类图书计出版及自然教育。我们相信，唯有了解，我们才会关心；唯有关心，我们才会行动；唯有行动，生命才有希望IBE 创立的宗旨是，用影像记录中国的生物多样性，传播中国的自然之美。我们专注于中国，深入野外第一线进行客观真实的影像记录，在此过程中充分地尊重自然，力求将自然之美展现给公众。

截至 2018 年，IBE 已经走过中国 17 个省、自治区的 60 个地区，开展了 94 次影像调查，先后参与的人达到 400 多人次，累计野外工作时间 8000 多天，行程 30 多万千米。IBE 的摄影师共拍摄了 8000 多个物种 30 多万张图片、10 多部自然纪录片和 3000 多小时的视频。IBE 开展了近百场讲座，我们的工作内容、摄影和丰富的项目成果先后被国内外 100 多家媒体报道。

Imaging Biodiversity Expedition (IBE) is a natural imaging institution founded by profession photographers in 2009. Since the very beginning, IBE has adopted an innovative mode of biologic survey: After each expedition, photos of different species taken by different professional natu photographers are collected as scientific evidence for biodiversity research.

IBE specializes in imaging biodiversity expedition, shooting & production of wildlife documentarie natural image exhibition, design & publishing of nature books, and nature education. We believe th understanding is the mother of care, care the mother of action, and action the mother of hope. IB is intended to spread the beauty of wild China and catalogue the biodiversity of our country throug images. With a sincere respect for nature, we venture deep into the wilderness for authentic ar objective imaging.

By 2018, IBE has been to 60 sites in 17 provinces/autonomous regions through 94 expedition during which the 400-odd participants have spent over 8,000 days in the wild and covered a distanc of more than 300,000km. We have, to our credit, over 300,000 photos of 8,000-odd species, dozen of wildlife documentaries, and more than 3,000 hours of videos. We have also held nearly 10 lectures. Our work, photographers and achievements have been successively covered by over 10 media at home and abroad.

朱仁斌 Zhu Renbin

高校教师

University Teacher

王　斌（怒江） Wang Bin (Nu River)

怒江自然保护区工作人员

Staff at Nujiang Nature Reserve

张兴伟 Zhang Xingwei

摄影爱好者

Amateur Photographer

王　斌 Wang Bin

自由职业者

Freelancer

左凌仁 Zuo Lingren

公司职员

Clerk

程　斌 Cheng Bin

职业摄影师

Professional Photographer

刘彦鸣 Liu Yanming

平面广告摄影师

Print Ad Photographer

崔　林 Cui Lin

平面设计师

Graphic Designer

刘思远 Liu Siyuan

旅行探险家

Adventurer

雷　波 Lei Bo

自由职业者

Freelancer

牛　洋 Niu Yang
助理研究员
Research Assistant

吴　双 Wu Shuang
高级工程师
Senior Engineer

吴秀山 Wu Xiushan
高级兽医
Senior Veterinarian

袁明辉 Yuan Minghui
职业摄影师
Professional Photographer

范　毅 Fan Yi
职业摄影师
Professional Photographer

黄一峰 Huang Yifeng
生态设计师
Eco-Designer

向定乾 Xiang Dingqian
保护区工作人员
Staff at a Nature Reserve

李　成 Li Cheng
工业设计师
Industrial Designer

徐廷程 Xu Tingcheng
医生
Doctor

侯　勉 Hou Mian
高校教师
University Teacher

张　瑜 Zhang Yu
杂志编辑
Magazine Editor

朱浩文 Zhu Haowen
博士生
Ph. D.

郭　亮 Guo Liang

职业摄影师

Professional Photographer

秦大公 Qin Dagong

工程师

Engineer

徐　健 Xu Jian

职业摄影师

Professional Photographer

彭建生 Peng Jiansheng

企业管理者

Company Manager

董　磊 Dong Lei

高校教师

University Teacher

吴立新 Wu Lixin

水下摄影师

Underwater Photographer

2009—2018
IBE 影像生物多样性调查地区

2009 年　广西西南崇左石灰岩地区
2009 年　云南梅里雪山国家公园
2009 年　广西桂林雁山园

2010 年　云南丽江老君山国家公园
2010 年　云南西双版纳尚勇保护区
2010 年　四川石渠地区
2010 年　广西大明山
2010 年　西藏林芝雅鲁藏布大峡谷景区

2011 年　陕西秦岭长青国家级自然保护区
2011 年　西藏林芝巴松措景区

2012 年　云南普洱太阳河国家公园
2012 年　云南孟连那勒地区
2012 年　西藏鲁朗花海景区
2012 年　青海三江源之长江源地区
2012 年　云南普洱西盟永不落地区
2012 年　河北唐山曹妃甸湿地

2013 年　贵州大娄山地区
2013 年　西藏珠峰南坡陈塘沟地区
2013 年　西藏羌塘地区
2013 年　云南怒江高黎贡山国家级自然保护区
2013 年　青海三江源之黄河源地区
2013 年　西藏阿里神山圣湖地区
2013 年　吉林汪清东北虎栖息地
2013 年　西藏墨脱地区

2014 年　四川老君山国家级自然保护区
2014 年　西藏珠峰雪豹栖息地
2014 年　青海三江源之澜沧江源地区
2014 年　山西临汾涝洰河地区
2014 年　广东珠海伶仃洋海岛
2014 年　四川卧龙国家级自然保护区
2014 年　西藏山南地区

2015 年　辽宁丹东鸭绿江口湿地
2015 年　辽宁医巫闾山国家级自然保护区
2015 年　广西扶绥那坡地区
2015 年　云南普洱太阳河国家级自然保护区

2016 年　河南郑州中牟地区
2016 年　内蒙古阿拉善左旗地区
2016 年　北京奥林匹克森林公园

2017 年　北京门头沟军庄镇地区
2017 年　江苏昆山市
2017 年　黑龙江黑河胜山国家级自然保护区

2018 年　云南普洱糯扎渡自然保护区

后记

志者，记录也；影像志者，以照片的方式记录事物也；自然影像志者，以实地拍摄的照片记录自然界中动植物之状态也；中国自然影像志者，以实地拍摄的照片记录中国自然界之动植物也。中国的动植物种类繁多，不可能置于一本书之中，故眼前的这本《中国自然影像志》只能有选择地包括中国之珍稀动植物及其生存的环境。

这本影像志是一个年轻团队的作品，他们是一群摄影师，又是一群博物学家，即背着照相机奔波于高山大泽中的博物学家。他们有 30 多个人，其中最年长者 50 岁，最年幼者 20 岁，从事着编辑、记者、教师、作家、学者、医生、公务员、环保工作者等不同的职业，为了一个共同的目标组成了一个专业的、成熟的、充满激情的团体。他们以相机为工具，以感官和心灵为媒介，以保护生态为宗旨，以在远离城市的地方的艰苦劳作向大众传播真、美和感情，是把生物多样性调查作为一项事业并在其中投入巨大热情的带照相机的博物学家。自 2010 年 9 月影像生物调查所（IBE）正式成立以来，他们的足迹遍及半个中国，行程 30 多万千米，开展了 94 次影像调查，拍摄了 8000 多个物种、30 多万张图片。可以说，这本《中国自然影像志》中的图片是优中之优。

博物学是一门古老的、曾经辉煌过的学科，它的最本质的要求是亲力亲为，即认识一件事物一定要通过自己的眼睛去观察、通过自己的耳朵去聆听、通过自己的舌尖去品尝、通过自己的手去触摸、通过自己的脚去丈量，所以博物学是一门需要参与者亲自动手、动脑、动心的学问。这些年轻的博物学家打起行囊，背着相机，走出城市，深入荒野，到人迹罕至的地方去捕捉人们难得一见的影像，在完成调查项目的同时把一帧帧真实而精美的图片呈现在世人面前。他们背着沉重的器材，无惧蚊虫的叮咬，经风雨，涉急流，攀悬崖，睡帐篷，饥一顿饱一顿，或者一动不动地等待数小时，甚至几天、几星期，为的就是一只动物或一株植物，希望亲手拍一张角度和构图都具有创意的照片。然而，这一张张照片表现的不仅仅是“多识于草木鸟兽之名”，还有它们的栖息地，它们繁殖、迁徙的地方，总之是不加任何伪装或修饰的大自然。动植物与其环境之间的适应与互动，才是生物多样性的真谛——蹚过西水河的大熊猫，重返荒野的普氏野马，绿浪来袭的大紫胸鹦鹉，笑傲群山的黑鹳，飞越山巅的黑颈鹤，晨昏之光中的藏羚羊，“巨人”塔黄，“风中之火”的红花绿绒蒿，变色蜥蜴的“瑜伽术”，等等。它们或雄踞于高山，或藏身于荒原，或嬉戏于悬崖，或隐匿于水下，或翱翔于蓝天，或潜伏于黑夜，或摇摆于风中，或躲避于巨石之间，无一不是生物与环境之间的适应与和谐。这些年轻的博物学家是一些有信仰的理想主义者，相信梭罗的话："荒野蕴含着人类的希望。"也相信庄子的话："天地与我并生，而万物与我为一。"

"天人合一"，或者"物我为一"，是中国人自然观的基础与核心，在其漫长的历史发展过程中，虽然有不同的重点，也出现过不同的解读，但是万变不离其宗，始终守住一个"一"字。例如，老子认为，万物源于自然且平等相处，"道生一，一生二，二生三，三生万物，万物负阴而抱阳，冲气以为和。""道大，天大，地大，人亦大。域中有四大，而人居其一焉。"孔子说："天何言哉？四时行焉，万物生焉，天何言哉。""仁者乐山，智者乐水。"秦汉时期，董仲舒明确指出："天人之际，合而为一，分为阴阳，判为四时，列为五行。"自此，"天人感应"之说出焉。魏晋时期，"庄老告退，而山水方滋"，以自然山水评价人物，自然与人之间相融相洽的关系更加彰显。当然，其中也有荀子一派，提出"制天命而用之"的观点，唐代柳宗元、刘禹锡以"各行不相预""天人交相胜"相继，形成了与主流意识不同的派别。自唐以降，直至民国，"天人合一"或"物我为一"的观念似乎占了统治地位。说到当下，《易传》的"天人协调"说更合乎今人的口味。《周易大传》主张"裁成天地之道，辅相天地之宜"，"范围天地之化而不过，曲成万物而不遗"，这样的观念潜在于人们的意识之中，或隐或显地支配着人们改善环境、保护生态的活动。

"天人合一"，或者"物我为一"，顾名思义，是说人和天是一个整体，声息相通，同心同体，形成了一个不可分割的共同体。人是小宇宙，天是大宇宙，人心即是宇宙，天即是自然，自然即是万物，人是其中之一，并无特殊的地位。人与万物的关系是一与多的关系，只有和谐相处才是唯一的出路。天人为一，天人相连，天人感应，天人合德。就拿动物与人的关系来说，人的同情与爱护及于动物，反过来，动物也在人的想象中将其同情与爱护及于人类。人与动物之间的互动和交流，才能形成一个和谐的世界。在这方面，唐代的大诗人白居易可以说是一个代表。他写有以《鸟》为题的诗："谁道群生性命微，一般骨肉一般皮。劝君莫打枝头鸟，子在巢中盼母归。"拳拳爱惜之情溢于言表。他又写有《赎鸡》："……常慕古人道，仁信及鱼豚。见兹生恻隐，赎放双林园。开笼解索时，鸡鸡听我言：与尔镪三百，小惠何足论。莫学衔环雀，崎岖谩报恩。"付出而不求回报的态度跃然纸上。他还写有《放旅雁》："九江十年冬大雪，江水生冰树折枝。百鸟无食东西飞，中有旅雁声最饥。雪中啄草冰上宿，翅冷腾空飞动迟。江童持网捕将去，手携入市生卖之。我本北人今谴谪，人鸟虽殊同是客。见此客鸟伤客人，赎汝放汝飞入云。雁雁汝飞向何处？第一莫飞西北去。淮西有贼讨未平，百万甲兵久屯聚。官军贼军相守老，食尽兵穷将及汝。健儿饥饿射汝吃，拔汝翅翎为箭羽。"一种悲天悯人（鸟）的仁爱之言声声在耳。在诗人的千叮万嘱中，人们似乎听见了"鸡鸡"与"雁雁"的鸣叫，它们感谢诗人的关切与担心。交流与互动，是和谐世界的必要条件。

这些背着相机的博物学家向我们展示的是大自然的真，大自然的实，大自然的美。常言道：眼见为实。但是，眼见而没有准确的细节和氛围，无非是一个个孤零零、干巴巴的没有生命的动物或植物，还算不得完全的真、实和美，必须有细节的烘托和环境的映衬才能显示物体的形态、样貌和精神。这里的每一张照片都有呼吸，有动作，有生命。在这些年轻的博物学家的眼中，鸟兽鱼虫，草木山川，都有一个与人类平等的生命，他们不需要让这些活泼的生命体摆出各种姿态，或者人为地造出某种情势。他们要的是原生态。博物学不仅提供真实，还提供美感；不仅提供知识，还提供欣赏的对象。化瞬间为永恒，是美的真谛。有了真与实的自然，有了自然的美感，也许还不够，还要有参与的热情，即拍照的时候有感情投入、有发自内心的快乐和对每一种生物由衷的喜爱。博物学既是科学，又是艺术，还是感情。它的对象集中了知识、美感和激情，是一种人人可以参与、可以分享的学问。摄影师和每一只动物、每一株植物接触的时候，都怀有一种爱护、保护甚至守护的感情，他们之间用眼神、手势和面部的表情进行着不同层面的沟通与交流。面对一张张照片，人们可以感受到摄影师的快乐、喜悦，甚至骄傲。

IBE 的摄影师们无愧于博物学家的身份，他们满怀激情，用他们的成绩向我们表明大自然的每一种生物都是真实的，都是美的，都有其存在的价值。他们的行为证明了：苦中有乐，乐中有知，知中有物，物中有情，情中有像，像中有真，真中有美，美中有爱。真、美、爱，这是博物学的三元素，也是 IBE 的博物学家们矢志不渝的信念。

是为序。

郭宏安

中国社会科学院荣誉学部委员

2018 年 6 月，北京

Afterword

Nature and Wildlife Photography in China is, as the title suggests, an album comprising snapshots of animals and plants in wild China. Given the sheer variety of wildlife, the album collects only the pictures of rare animals and plants as well as their living environment.

The album is a lucubration of 30-odd naturalist photographers aged 20-50 .Working as doctors, editors, environmentalists, journalists, scholars, teachers, writers and civil servants in most of the time, the shutter bugs were brought together by a shared commitment to biodiversity expedition. Remote mountains and lakes bear witness to their passion, professionalism, and unremitting endeavor to spread beauty, love and truth to the public. Since the Imaging Biodiversity Expedition (IBE) was officially founded in September 2010, the team has crossed half the country through 60 expeditions which cover a distance of 300,000km, and collected 300,000 photos of 8,000 species. Photos included in the album are the crème de la crème.

Naturology is a time-honored, once-flourishing science which calls on explorers to see with their own eyes, hear with their own ears, taste with their own tongues, touch with their own hands, measure with their own tools, think with their own minds and feel with their own hearts. In answer to the call, the young naturalists said farewell to the cities and set out for the wilderness, aiming to present authentic, beautiful images to readers through successful expeditions. With a heavy lens on the back, they forded rapids and scaled cliffs, enduring the elements, hunger and bug bites. Often times they had to wait motionless for hours, days or even weeks before a specific animal showed up; oftentimes they had to cast about for an original angle to fit a certain plant into an original background. The photos they were about to take should be the real face of not only the wildlife, but also its living space, breeding site or migration route. Wildlife's adaptation to and interaction with the environment is the true meaning of biodiversity, as evidenced by the panda wading through Xishui River, the Przewalski's horses returning to the wild, the Derbyan parakeets rebounding from near extinction, the black stork overlooking the mountains, the black-necked crane overflying the peak, the Tibetan antelope cavorting in the twilight, the towering Rheum nobile, the fiery Meconopsis punicea, and the "yoga-savvy" chameleon. The call of the wild echoes through time and space. As Henry David Thoreau held that in wildness is the preservation of the world, so did Zhuangzi argue that Heaven, Earth, and I were produced together, and all things and I are one.

"Heaven-man oneness", or "unity of Universe and self", is the be-all and end-all of the Chinese view of nature. The passage of time has seen it evolve into different schools of thoughts, yet all of them are centered around the keywords of "oneness" and "unity". "The Tao produced One; One produced Two; Two produced Three; Three produced All things. All things leave behind them the Obscurity (out of which they have come), and go forward to embrace the Brightness (into which they have emerged), while they are harmonized by the Breath of Vacancy……The Tao is great; Heaven is great; Earth is great; and the (sage) king is also great. In the Universe there are four that are great, and the (sage) king is one of them," said Laozi."Heaven does not speak in words. It speaks through the rotation of the four seasons and the growth of all living things...The wise enjoy water; the humane enjoy mountains," said the Confucius."The Qi of Heaven and that of Earth merge into one; it evolves into yin and yang, the four seasons, and the five elements of metal, wood, water, fire, and earth," said Dong Zhongshu, a Han Dynasty scholar who developed the theory of "Heaven-man induction". During the Wei and Jin Dynasties, "the ideology of Laozi and Zhuangzi had receded into the background, and the theme of mountains and waters then began to flourish". The practice of associating character with mountains and waters further highlighted the harmony between man and nature. Minor conceptions, which range from "subjugate the course of Heaven and appropriate it (for human ends)", advanced by Xunzi, to"(the matters of) Heaven and man are carried out separately without interfering with each other" and "Heaven and man triumph over each other", propounded by Liu Zongyuan and Liu Yuxi, two Tang Dynasty poets, respectively, seem to be flawed in hindsight. From the Tang Dynasty to the Republic Period, "Heaven-man oneness", or "unity of Universe and self", appeared to hold a paramount sway. Today, "Heaven-man coordination", a doctrine proposed by *The I Ching*, has got its second wind. Our subconscious identification with such ideas as "the ruler divides and completes the course of Heaven and Earth" and "the spirit molds and encompasses all transformations of Heaven and Earth without mistake, and it stoops to being things into completion without missing any" spurs us into environmental campaigns.

"Heaven-man oneness", or "unity of Universe and self", means that Heaven and man are linked closely with each other, and together they constitute an inseparable whole. Every human mind is a universe within the Universe, and every man is a common part of nature. In the imaginative eyes of Bai Juyi, a renowned Tang Dynasty poet, animals indeed communicate with us and share our virtue:

....

The Bird

Who said all living creatures' lives were not worth mentioning
Equal flesh and blood, same skin
Advise you not to fowl the bird on the branch
Its children in the nest thirst for their mother back

....

Ransoming the Chicks

Oftentimes do I admire the sages of days olden
Whose admonition and benevolence reached ponds and pens.
The very sight of you aroused my compassion
And determination to bring you back to the gardens.
The cage opened, the chains untied,
Chicks, please bear my words in mind:
Three hundred coins for a ransom -
Too small to mention a favor.
Never emulate the legendary sparrow,
Who took pains to repay his savior.

....

Setting a Migrant Goose Free

Snows heavy in Xunyang this tenth-year winter,
River water spawns ice, tree brunches break and fall;
And hungry birds flock east and west by the hundred,
A migrant goose crying starvation loudest among them.
Pecking through snow for grass, sleeping nights on ice,
Its cold wings lumber slower and slower up into flight,
And soon it's tangled in a river-boy's net, carried away
Snug in his arms, and put for sale alive in the market.
Once a man of the north, I'm accused and exiled here.
Man and bird: though different, we're both visitors,
And it hurts a visiting man to see a visiting bird's pain,
So I pay the ransom and set you free. Goose, o soaring

Goose rising into the clouds - where will you fly now?

Don't fly northwest: that's the last place you should go.

There in Huaixi, rebels still loose, there's no peace,

Just a million armored soldiers long massed for battle:

Imperial and rebel armies grown old facing each other.

Starved and exhausted -they'd love to get hold of you,

Those tough soldiers. They'd shoot you and have a feast,

Then pluck your wings clean to feather their arrows.

…………

Interaction between man and nature is the sine qua non for a harmonious world. Aren't the birds chirping, gobbling and squawking their thanks for the poet's care and concern?

The true beauty of nature is what the young naturalists try to unfold. A creature is not truly beautiful unless its appearance, shape and spirit are clearly presented against a specific background. Every photo in the album depicts a living, breathing soul. In the naturalists' eyes, all animals and plants are equal to man. Instead of forcing a pose or creating a scene, they want their "models" to be as natural as possible. Naturology is no less a science of beauty than one of truth, yet it takes more than both to make a perfect picture, the ultimate art of transforming an instant into eternity. A naturalist photographer should be the patron saint of wildlife. He should be in a position to communicate with his models by gestures, eye contact and facial expressions. Each of his work should be the embodiment of his affection, enthusiasm, joy, and even pride.

IBE's team are worthy of the title of naturalist photographers. They are able to enjoy hardship and learn through joy, appreciate the true beauty of wildlife, and channel their love of nature into photography. Their emotionally charged works prove that every creature is truly beautiful and has its own value. Beauty, love and truth, the three elements of naturology, are bound to be their unswerving pursuits.

Guo Hong'an

Honorary Member of the Chinese Academy of Social Sciences

Beijing, June 2018